ENERGY SCIENCE, ENGINEERING AND TECHNOLOGY

U.S. WIND POWER INDUSTRY MARKET: CONDITIONS AND TRENDS

ENERGY SCIENCE, ENGINEERING AND TECHNOLOGY

Additional books in this series can be found on Nova's website under the Series tab.

Additional E-books in this series can be found on Nova's website under the E-books tab.

ENERGY SCIENCE, ENGINEERING AND TECHNOLOGY

U.S. WIND POWER INDUSTRY MARKET: CONDITIONS AND TRENDS

LIAM E. DEMPSEY
EDITOR

Nova Science Publishers, Inc.
New York

For permission to use material from this book please contact us:
Telephone 631-231-7269; Fax 631-231-8175
Web Site: http://www.novapublishers.com

Additional color graphics may be available in the e-book version of this book.

LIBRARY OF CONGRESS CATALOGING-IN-PUBLICATION DATA

U.S. wind power industry market : conditions and trends / editor, Liam E. Dempsey.
p. cm.
Includes index.
ISBN 978-1-61209-559-2 (hardcover)
1. Wind power industry--United States. 2. Wind power--United States. I. Dempsey, Liam E. II. Title: US wind power industry market conditions and trends.
HD9502.5.W553U548 2011
333.9'20973--dc22

2011003979

Published by Nova Science Publishers, Inc. † New York

CONTENTS

PREFACE

Despite grim predictions at the close of 2008, the U.S. wind power industry experienced yet another record year in 2009, once again surpassing even optimistic growth projections from years past. At the same time, the combination of the financial crisis and lower wholesale electricity prices has taken a toll on the wind power industry, dampening expectations for 2010. This new book examines the U.S. wind power industry market conditions and trends, and analyzes the impact of wind energy installations on domestic manufacturing and trade.

Chapter 1- Despite grim predictions at the close of 2008, the U.S. wind power industry experienced yet another record year in 2009, once again surpassing even optimistic growth projections from years past. At the same time, the combination of the financial crisis and lower wholesale electricity prices has taken a toll on the wind power industry, dampening expectations for 2010. Key findings from this year's "Wind Technologies Market Report" include:

Chapter 2- The Obama administration has prioritized expansion of renewable energy as a way to reduce greenhouse gas emissions and revitalize the economy. Central to this strategy is investment in wind energy, which is already adding thousands of megawatts (MW) each year to the U.S. electric grid. Yet there is debate about the economic impact of wind energy investment and the extent to which the U.S. industry will benefit from the growth in wind turbine installations. This paper thus analyzes the impact of the increase in wind turbine demand from 2005 to 2009 on U.S. production and trade.

Chapter 3- U.S. wind turbine manufacturing and trade are rapidly expanding due to growing domestic demand for wind turbines. Annual U.S. wind turbine installations[1] increased from 1,672 to 8,545 megawatts (MW) (over 400 percent) from 2003 to 2008, leading to a growth in imports and rising investment in U.S. production by American, European, and Asian companies.[2] New U.S. manufacturing plants are primarily intended to meet domestic demand, but some U.S. manufacturers are beginning to examine export opportunities or open overseas manufacturing plants.

In: U.S. Wind Power Industry Market Conditions and Trends ISBN: 978-1-61209-559-2
Editor: Liam E. Dempsey

Chapter 1

2009 WIND TECHNOLOGIES MARKET REPORT

U.S. Department of Energy

ACKNOWLEDGMENTS

For their support of this project, the authors thank Jacques Beaudry-Losique, Megan McCluer, Jim Ahlgrimm, Michele Desautels, and Patrick Gilman of the U.S. Department of Energy's (DOE) Wind & Water Power Program. For reviewing elements of this paper, we thank: Michael Goggin, Ron Stimmel, Liz Salerno, Jessica Isaacs, and Kathy Belyeu (AWEA); Brendan Kirby (consultant); Charlie Smith (UWIG); Jason Gifford and Bob Grace (Sustainable Energy Advantage); Andrew David (U.S. International Trade Commission); Jacques Beaudry-Losique, Liz Hartman, Jim Ahlgrimm, Brian Naughton, Patrick Gilman, and Ron Harris (DOE); Roger Hill (Sandia National Laboratory); Bob Thresher and Maureen Hand (NREL); James Browning (BCS, Inc.); Chris Namovicz (Energy Information Administration); Jennifer States (PNNL); Adam Hanna and Ric O'Connell (Black & Veatch); Jim Walker (enXco); Doug Larson (Western Governors' Association); Ed DeMeo (Renewable Energy Consulting Services, Inc.); Mike O'Sullivan (NextEra Energy Resources); Steve Clemmer and Alan Nogee (Union of Concerned Scientists). Special thanks to the American Wind Energy Association (Kathy Belyeu and Elizabeth Salerno) for the use of their database of wind power projects, and for providing other data as discussed in the Appendix. We also thank Andrew David (U.S. International Trade Commission), Frank Oteri (NREL), and Billy Roberts (NREL) for their contributions to the wind equipment manufacturing sections; Walt Musial (NREL) as well as Bonnie Ram and Wendy Wallace (Energetics) for their assistance with the offshore wind energy discussion; Donna Heimiller (NREL) for the wind power project map; and Michelle Kubik and Kathleen O'Dell (NREL) for assistance with layout, formatting, and production. Berkeley Lab's contributions to this report were funded by the Wind & Water Power Program, Office of Energy Efficiency and Renewable Energy of the U.S. Department of Energy under Contract No. DE-AC02-05CH1 1231. The authors are solely responsible for any omissions or errors contained herein.

LIST OF ACRONYMS

AWEA	American Wind Energy Association	MMS	Minerals Management Service
BPA	Bonneville Power Administration	MNDOC	Minnesota Department of Commerce
CCGT	combined cycle gas turbine	MSR	Modesto-Santa Clara-Redding Public Power Agency
COD	commercial operation date	MW	megawatt
CREB	clean renewable energy bond	MWh	megawatt-hour
CREZ	competitive renewable energy zone	NERC	North American Electric Reliability Corporation
DOE	U.S. Department of Energy	NREL	National Renewable Energy Laboratory
EIA	Energy Information Administration	OEM	original equipment manufacturer
ERCOT	Electric Reliability Council of Texas	O&M	operations and maintenance
EWITS	Eastern Wind Integration and Transmission Study	POU	publicly owned utility
FAA	Federal Aviation Administration	PPA	power purchase agreement
FERC	Federal Energy Regulatory Commission	PTC	Production Tax Credit
GE	General Electric Corporation	PUC	public utility commission
GW	gigawatt	PUCT	Public Utility Commission of Texas
G&T	generation and transmission	REAP	Rural Energy for America Program
HTS	Harmonized Tariff Schedule	REC	renewable energy certificate
ICE	IntercontinentalExchange	RFI	request for information
IOU	investor-owned utility	RPS	renewables portfolio standard
IPP	independent power producer	RTO	regional transmission organization
IREC	Interstate Renewable Energy Council	SPP	Southwest Power Pool
IRS	Internal Revenue Service	TVA	Tennessee Valley Authority
ISO	independent system operator	USDA	United States Department of Agriculture
ITC	investment tax credit	USITC	U.S. International Trade Commission
kW	kilowatt	UWIG	Utility Wind Integration Group
kWh	kilowatt-hour	WAPA	Western Area Power Administration
LADWP	Los Angeles Department of Water& Power	WWSIS	Western Wind and Solar Integration Study
LIBOR	London Interbank Offered Rate		
MISO	Midwest Independent System Operator		

EXECUTIVE SUMMARY

Despite grim predictions at the close of 2008, the U.S. wind power industry experienced yet another record year in 2009, once again surpassing even optimistic growth projections from years past. At the same time, the combination of the financial crisis and lower wholesale electricity prices has taken a toll on the wind power industry, dampening expectations for 2010. Key findings from this year's "Wind Technologies Market Report" include:

- **Wind Power Additions in 2009 Shattered Old Records, with roughly 10 GW of New Capacity Added in the United States and $21 Billion Invested.** The pace of utility-scale wind power capacity additions in 2009 was 20% higher than the previous U.S. record set in 2008, while cumulative wind power capacity grew by 40%. This was achieved despite the financial crisis that roiled the wind power industry in 2009, and the significant reductions in wholesale electricity prices that began in mid- to late-2008 and have continued to the present. A variety of market drivers allowed year-on-year installation growth to persist in 2009, including: carryover of projects initially planned for completion in 2008; elements of the *American Recovery and Reinvestment Act of 2009* (Recovery Act), including the Section 1603 Treasury Grant Program; the expiration of bonus depreciation rules at the end of 2009; and state renewables portfolio standards.

- **Wind Power Contributed 39% of All New U.S. Electric Generating Capacity in 2009.** This is down from 44% in 2008, but exceeds wind power's contribution of 35% in 2007, 18% in 2006, 12% in 2005, and less than 4% from 2000 through 2004. For the fifth consecutive year, wind power was the second-largest new resource added to the U.S. electrical grid in terms of nameplate capacity, behind natural gas plants, but ahead of new coal power.
- **The United States Continued to Lead the World in Cumulative Wind Power Capacity, but Was Overtaken by China in Annual Additions.** After four years of leading the world in annual wind power capacity additions, the United States dropped to second place in 2009, capturing roughly 26% of the worldwide market (behind China's 36% market share). At the end of 2009, cumulative wind power capacity in the United States stood at more than 35,000 MW, ahead of China's 25,853 MW and Germany's 25,813 MW. Several countries are beginning to achieve relatively high levels of wind energy penetration in their electricity grids: end-of-2009 wind power capacity is projected to supply the equivalent of roughly 20% of Denmark's electricity demand, 14% of Spain's and Portugal's, 11% of Ireland's, and 8% of Germany's. In the United States, the cumulative wind power capacity installed at the end of 2009 would, in an average year, be able to supply roughly 2.5% of the nation's electricity consumption.
- **Texas Achieved Higher Annual Capacity Additions than Other States, While Four States Have Surpassed 10% Wind Energy Penetration.** With 2,292 MW installed in 2009 alone, Texas dominated the 28 other states in which new large-scale wind turbines were installed in 2009 (the next highest were Indiana with 905 MW and Iowa with 879 MW). In terms of estimated wind energy supply as a proportion of in-state electricity generation, the front-runners include Iowa (19.7%), South Dakota (13.3%), North Dakota (11.9%), and Minnesota (10.7%). Some utilities are seeing higher percentages of wind energy supply than these state totals, with nine utilities estimated to have in excess of 10% wind energy on their systems.
- **Offshore Wind Power Project and Policy Developments Accelerated in 2009.** To date, all wind power installations in the United States have been located on land, but there is also interest in offshore wind power development and 2,476 MW of offshore projects have advanced significantly in the permitting and development process. Of those projects, three have signed or proposed power purchase agreements with terms and details have been made public. Notably, after nine years in the permitting process, the Cape Wind project was granted approval by the Department of Interior

in April 2010, and a variety of other recent project and policy announcements demonstrate accelerated activity in the offshore wind energy sector.

- **Data from Interconnection Queues Demonstrate that an Enormous Amount of Wind Power Capacity Is Under Consideration.** At the end of 2009, there were roughly 300 GW of wind power capacity within the transmission interconnection queues administered by independent system operators, regional transmission organizations, and utilities reviewed for this report – nearly nine times the installed wind power capacity. This wind power capacity represented almost 60% of all generating capacity within these queues at that time, and was nearly three times as much capacity as the next-largest resource (natural gas). Most (93%) of this wind power capacity is planned for the Midwest, Mountain, Texas, PJM, SPP, and Northwest regions. Not all of this capacity will ultimately be built as planned, but these data demonstrate the high level of developer interest in wind power.
- **GE Remained the Top Turbine Manufacturer in the United States Market, but a Growing Number of Other Manufacturers Are Capturing Market Share.** GE secured 40% of U.S. market share (by capacity) in 2009, followed by Vestas (15%), Siemens (12%), Mitsubishi (8%), Suzlon (7%), Clipper and Gamesa (6% each), REpower (3%), Acciona (2%), and Nordex (1%). Manufacturers with modern wind turbines installed in the United States now hail from not just the United States, Europe, and Japan, but also from India and, for the first time in 2009, China. In 2009, U.S.-owned GE was the second-leading supplier of turbines *globally*, while Clipper was the 13th largest global supplier. On a worldwide basis, perhaps the most significant story of 2009 was the growing market share of Chinese turbine manufacturers; to date, that growth has been based almost entirely on sales to the Chinese market, but Chinese manufacturers began to express strong interest in the U.S. market in 2009.
- **Domestic Wind Turbine and Component Manufacturing Investments Remained Strong in 2009, but the Financial Crisis and Weak Turbine Sales Slowed the Sector's Growth.** Seven of the ten wind turbine manufacturers with the largest share of the U.S. market in 2009 now have one or more manufacturing facilities operating in the United States, and two of the remaining three have announced specific plans to open facilities in the future. These figures compare to just one utility-scale wind turbine manufacturer (GE) assembling nacelles in the United States in 2004. In addition, a considerable number of new component manufacturing facilities were either announced or opened in 2009, by both foreign and domestic firms. Nevertheless, weak demand for new wind turbine orders and the poor state of the U.S. economy led to a net loss of 1,500 wind turbine and component manufacturing jobs in 2009, according to the American Wind Energy Association (AWEA). As a result, AWEA estimates that overall U.S. employment in the wind energy sector held steady at 85,000 full- time jobs in 2009; of these, 18,500 are estimated by AWEA to be turbine and component manufacturing jobs.
- **A Growing Percentage of the Equipment Used in U.S. Wind Power Projects Has Been Sourced Domestically in Recent Years.** U.S. trade data show that the United States remained a large importer of wind power equipment in 2009, but that wind power capacity growth has outpaced the growth in imports in recent years. As a result, a growing amount of the equipment used in wind power projects is being

sourced domestically as domestic and foreign companies seek to minimize transportation costs and currency risks by establishing local manufacturing capabilities. Imports of wind turbines and select components in 2009 are estimated at $4.2 billion, down from $5.4 billion in 2008. When presented as a fraction of total equipment-related wind turbine costs, the overall import fraction is estimated to have declined from roughly 50% in 2008 to 40% in 2009 as domestic manufacturing investments outpaced import growth.

- **The Average Nameplate Capacity, Hub Height, and Rotor Diameter of Installed Wind Turbines Increased.** The average nameplate capacity of wind turbines installed in the United States in 2009 increased to roughly 1.74 MW, up from 1.66 MW in 2008 and 1.65 MW in 2007. Since 1998-99, average turbine nameplate capacity has increased by 145%, but growth in this metric has slowed in recent years due to the dominance of GE's 1.5 MW turbine and as a result of the logistical challenges associated with transporting larger turbines to project sites. In addition to nameplate capacity ratings, average hub heights and rotor diameters have also scaled with time, to 78.8 and 81.6 meters, respectively, in 2009. Since 1998-99, the average turbine hub height has increased by 40%, while the average rotor diameter has increased by 69%: these trends are one of several factors impacting the project- level capacity factors highlighted later.
- **The Average Size of Wind Power Projects Resumed its Upward Trend.** Wind power projects installed in 2009 averaged nearly 91 MW, which is below the 120 MW average size of projects built in 2007, but is otherwise larger than in any previous period. Larger project sizes reflect an increasingly mature energy source that is beginning to penetrate into the domestic electricity market in a significant way.
- **Consolidation Among Wind Project Developers Continues.** At least six significant acquisition or investment transactions involving roughly 18 GW of in-development wind power projects were announced in 2009, compared to the five transactions and 19 GW in 2008. This is well below the 11 transactions and 37 GW in 2007, and the 12 transactions and 34 GW in 2006. The more-subdued pace of activity since 2007 may be due to the fact that many of the prime targets for investment and/or acquisition were acquired in earlier years. In addition, some traditional buyers of wind assets may have decided to reign in new investments following aggressive purchases made in previous years, while some developers who might otherwise entertain offers may be holding out for better pricing as the market recovers. Looking ahead, the relatively weak demand for wind energy projected in 2010, coupled with an influx of cash from the Section 1603 Treasury grant program, may help to drive continued consolidation.
- **Treasury Cash Grant Expands Financing Options, Buoys the Wind Sector.** To reduce the market's dependence on scarce and costly third-party tax equity, Section 1603 of the Recovery Act enables wind power projects to temporarily choose a 30% cash grant administered by the U.S. Treasury in lieu of either the production tax credit (PTC) or a 30% investment tax credit (ITC). Owners of more than 6,400 MW of the wind power capacity installed in 2009 elected the grant in lieu of the PTC, and as much as 2,400 MW of this capacity may not have been built in 2009 had the cash grant not been available. Only about seven of the more-than-sixty 2009 projects that

elected the grant were financed using third- party tax equity; many of the rest substituted project-level term debt – which became increasingly available as 2009 progressed – in place of third-party tax equity.

- **Private IPP Project Ownership Remained Dominant, but Utility Ownership Increased.** Private independent power producers (IPPs) own 83% of all new wind power capacity installed in the United States in 2009, and also 83% of cumulative capacity. In a continuation of the trend begun several years ago, however, 16% of total wind power additions in 2009 are owned by electric utilities, who now own 15% of the cumulative wind power capacity in the United States. Community wind power projects account for the remaining 2% of both annual and cumulative capacity.
- **Long-Term Contracted Sales to Utilities Remained the Most Common Sales Arrangement, but Merchant Plants Were Surprisingly Abundant in 2009.** Investor- owned utilities continued to be significant purchasers of wind power, with 36% of the new 2009 capacity and 44% of cumulative capacity selling power to these utilities under longterm contract. Publicly owned utilities purchased another 22% and 18%, respectively. Surprisingly, given the tightening of credit requirements in the wake of the financial crisis, as well as sharply lower wholesale electricity prices, merchant/quasi-merchant projects were abundant in 2009, accounting for 38% of all new capacity and 26% of the cumulative capacity. It is possible that many of these merchant projects may now be seeking longer- term power purchase contracts in order to gain increased revenue stability.
- **Upward Pressure on Wind Power Prices Continued in 2009.** Although some of the cost pressures facing the industry in recent years have eased, it will take time before relief flows through the project development pipeline to impact overall average wind power prices. As such, 2009 was another year of rising wind power prices. The capacity-weighted average 2009 sales price for bundled power and renewable energy certificates, based on projects in the sample built in 2009, was roughly $61/MWh (in 2009 dollars), up from an average of $51/MWh for the sample of projects built in 2008, and nearly double the average of $32/MWh among projects built during the low point in 2002 and 2003. Among projects in the sample, those in Texas and the Heartland region have the lowest prices on average, while those in New England, California, and the East have the highest prices.
- **Sharp Drop in Wholesale Electricity Prices Makes the Near-Term Economics of Wind Energy More Challenging.** The increase in wind power prices in 2009, combined with the deep reduction in wholesale electricity prices (driven by lower natural gas prices), pushed wind energy from the bottom to the top of the wholesale electricity price range in 2009. Although low natural gas prices are, in part, attributable to the recession-induced drop in energy demand, the discovery and early development of significant shale gas deposits has reduced expectations for increases in natural gas prices going forward. As a result, natural gas prices may not rebound to earlier levels as the economy recovers, putting the near-term comparative economic position of wind energy at some risk.
- **The Installed Cost of Wind Power Projects Continued to Rise in 2009, but Reductions May Be on the Horizon.** Among a large sample of wind power projects installed in 2009, reported installed costs had a capacity-weighted average of

$2,120/kW. This average increased by $170/kW (9%) from the weighted-average cost of $1,950/kW for projects installed in 2008, and increased by $820/kW (63%) from the average cost of projects installed from 2001 through 2004. Installed costs may – on average – remain high for a period of time as developers continue to work their way through the dwindling backlog of turbines purchased in early 2008 at peak prices. There are expectations, however, that average costs will decline over time as the cost pressures (e.g., rising materials costs, the weak dollar, turbine and component shortages) that have challenged the industry in recent years ease. Differences in average installed costs among regions and by project size are also apparent in the data.

- **Wind Turbine Prices Have Begun to Show Signs of Easing, but Remain High By Historical Standards.** Since hitting a low point of roughly $700/kW in the 2000-2002 timeframe, average wind turbine prices have increased by approximately $800/kW (>100%) through 2009. Though turbine price increases have been the rule for a number of years, evidence is beginning to emerge that those days have ended, at least temporarily. Although visibility of turbine transaction prices declined in 2009 as the financial crisis took its toll and developers sat on turbine supply frame agreements that exceeded near-term development plans, cost pressures have eased since mid-2008. As a result, estimates of turbine price declines of as much as 15%, along with more favorable contract terms, have begun to emerge. These price reductions and improved terms can be expected, over time, to exert downward pressure on project costs and wind power prices.
- **Wind Project Performance Has Generally Improved Over Time, but Has Leveled Off in Recent Years.** Boosted primarily by higher hub heights and larger rotor diameters, cumulative sample-wide average capacity factors have, in general, gradually increased over time, from just over 24% in 1999 to a high of nearly 34% in 2008, before dropping back to 30% in 2009. The drop in 2009 is, in part, attributable to a relatively poor wind resource year in many parts of the country along with increasing amounts of wind power curtailment. Curtailment was particularly high in Texas (home to more than one-quarter of the nation's wind power capacity), with 17% of all potential wind energy generation within the Electric Reliability Council of Texas (ERCOT) curtailed in 2009. The sample-wide average U.S. capacity factor of 30% in 2009 would have reached 32% if not for the curtailment experienced in ERCOT and the Midwest. Other factors that may have slowed the rate of capacity factor increase for projects installed in more recent years include an enhanced emphasis on lower-quality wind resource sites (due to transmission and siting constraints), moderation of the increase in average hub heights and rotor diameters, and some challenges with turbine reliability.
- **Operations and Maintenance Costs Are Affected by the Age and Size of the Project, Among Other Factors.** Despite limited data availability, it appears that projects installed more recently have, on average, incurred lower O&M costs than older projects in their first couple of years in operation. Likewise, larger projects appear to experience lower O&M costs than do smaller projects, and O&M costs increase as projects age.
- **The Federal Policy Landscape Is Now More Favorable to Wind Energy than at Any Other Time in the Past Decade.** The Recovery Act of 2009 extended the PTC

for wind energy through 2012, and also implemented a number of other policy changes, including an option to elect a 30% cash grant or ITC in lieu of the PTC, the expansion and enhancement of a federal loan guarantee program managed by the U.S. Department of Energy (DOE), and a 30% tax credit for investment in advanced energy manufacturing facilities. In addition, $2.2 billion in new Clean Renewable Energy Bonds were allocated in 2009, and $60 million in U.S. Department of Agriculture (USDA) funding was distributed in the form of grants and loan guarantees, in part to fund wind power projects located in rural areas. Nonetheless, federal policy towards wind energy remains uncertain after 2012.

- **State Policies Play a Significant Role in Directing the Location and Amount of Wind Power Development.** From 1999 through 2009, 61% of the wind power capacity built in the United States was located in states with RPS policies; in 2009, this proportion was 57%. One new state (Kansas) established a mandatory RPS program in 2009, bringing the total to 29 states and Washington D.C. Utility resource planning requirements, voluntary customer demand for "green" power, state clean energy funds, and state and regional carbon reduction policies also play a role in supporting wind energy deployment.
- **Despite Progress on Overcoming Transmission Barriers, Constraints Remain.** Transmission development appears to be gaining some traction, but siting, planning, and cost allocation issues remain key barriers to transmission investment. In June 2010, the Federal Energy Regulatory Commission (FERC) issued a proposed transmission cost allocation rule aimed at easing planning and cost allocation barriers. States, grid operators, regional organizations, and the DOE continue to take proactive steps to encourage transmission investment to access remote renewable resources. Finally, progress was made in 2009 on a number of transmission projects that are designed, in part, to support wind power.
- **Integrating Wind Energy into Power Systems Is Manageable, but Not Free of Costs, and Market Operators Are Implementing Methods to Accommodate Increased Penetration.** Recent studies show that wind energy integration costs are below $10/MWh – and often below $5/MWh – for wind power capacity levels up to or exceeding 40% of the peak load of the system in which the wind power is delivered. Moreover, a number of strategies that can help to ease the integration of increasing amounts of wind energy – including the use of larger balancing areas, the use of wind forecasts, and intra-hour scheduling – are being implemented by grid operators across the United States.

In conclusion, 2009 continued a string of record-breaking years for the U.S. wind power industry. Looking ahead, expectations are for a slower year in 2010, due to a combination of the financial crisis, lower wholesale electricity prices, and lower demand for renewable energy. Wind power capacity additions in 2009 were buoyed, in part, by projects that were initially slated to be completed in 2008 but that carried over into 2009 when the PTC was extended, somewhat masking the underlying challenges facing the sector. With the extension of federal incentives through 2012, there is less motivation to complete projects in 2010 (though many projects will likely start construction in 2010 in order to be eligible for the 30% Treasury cash grant). Industry analysts project a range from 5,500 MW to 8,000 MW of wind power capacity likely to be installed in the United States in 2010, a drop of 20-45% compared to the nearly 10,000 MW installed in 2009. After a slower 2010, most predictions show market resurgence in 2011 and 2012, as the Recovery Act programs mature and as financing constraints ease. Beyond 2012, however, the picture is considerably less certain, due to the scheduled expiration of a number of federal policies at the end of that year, including the PTC, the ability to elect a 30% ITC in lieu of the PTC, and the ability to receive the 30% Treasury cash grant for projects that initiated construction by the end of 2010.

1. Introduction

The U.S. wind power industry experienced yet another record year in 2009, once again surpassing even optimistic growth projections from years past. At the same time, 2009 was a year of upheaval, with the global financial crisis impacting the wind power industry and with federal policy changes enacted to push the industry towards continued aggressive expansion. The year 2010, meanwhile, is anticipated to be one of some retrenchment, with expectations for fewer wind power capacity additions than seen in 2009.

The rapid pace of development and change within the industry has made it difficult to keep up with trends in the marketplace, yet the need for timely, objective information on the industry and its progress has never been greater. This report – the fourth in an ongoing annual series – attempts to meet this need by providing a detailed overview of developments and trends in the United States wind power market, with a particular focus on 2009.

As with previous editions, this report begins with an overview of key installation-related trends: trends in wind power capacity growth, how that growth compares to other countries and generation sources, the amount and percentage of wind energy in individual states and serving specific utilities, and the quantity of proposed wind power capacity in various interconnection queues in the United States. Next, the report covers an array of wind power industry trends, including developments in turbine manufacturer market share, manufacturing and supply-chain investments, wind turbine and wind power project size, project financing developments, and trends among wind power developers, project owners, and power purchasers. The report then turns to a discussion of wind project price, cost, and performance trends. In so doing, it reviews the prices paid for wind power in the United States, and how those prices compare to short-term wholesale electricity prices. It also describes trends in installed wind power project costs, wind turbine transaction prices, project performance, and operations and maintenance expenses. Next, the report examines other policy and market factors impacting the domestic wind power market, including federal and state policy drivers, transmission issues, and grid integration. Finally, the report concludes with a preview of possible near-term market developments.

This fourth edition updates data presented in the previous editions, while highlighting key trends and important new developments from 2009. New to this edition is a discussion of trends in the hub height and rotor diameter of wind turbines installed in the United States, new data on wind turbine and component imports into and exports from the United States, an expanded discussion of offshore wind energy development, and data on wind power curtailment. The importance of the *American Recovery and Reinvestment Act of 2009* (the Recovery Act) to wind energy in 2009 is reflected throughout the report. The report concentrates on larger-scale wind turbines, defined here as individual turbines or projects that *exceed* 100 kW in size.[1] The U.S. wind power sector is multifaceted, however, and also includes smaller, customer-sited wind turbines used to power residences, farms, and businesses. Data on these latter applications are not the focus of this report, though a brief discussion on *Small Wind Turbines* is provided on page 4.

Much of the data included in this report were compiled by Berkeley Lab, and come from a variety of sources, including the American Wind Energy Association (AWEA), the Energy Information Administration (EIA), and the Federal Energy Regulatory Commission (FERC). The Appendix provides a summary of the many data sources used in the report, and a list of specific references follows the Appendix. Data on 2009 wind power capacity additions in the United States are based on information provided by AWEA; methodological differences exist in the processing of those data, however, and the data presented here therefore varies somewhat relative to AWEA (2010a).[2] In other cases, the data shown here represent only a sample of actual wind power projects installed in the United States; furthermore, the data vary in quality. As such, emphasis should be placed on overall trends, rather than on individual data points. Finally, each section of this document focuses on historical market information, with an emphasis on 2009; with the exception of the final section, the report does not seek to forecast future trends.

2. Installation Trends

Wind Power Additions in 2009 Shattered Old Records, with roughly 10 GW of New Capacity Added in the United States and $21 Billion Invested

The U.S. wind power market delivered another record-shattering year in 2009, with 9,994 MW of new capacity added, bringing the cumulative total to more than 35,000 MW (Figure 1).[3] This growth translates into nearly $21 billion (real 2009 dollars) invested in wind power project installation in 2009, for a cumulative investment total of $66 billion since the beginning of the 1980s.[4]

Wind power installations in 2009 were not only the largest on record in the United States, but were 20% higher than the previous U.S. record, set in 2008. Cumulative wind power capacity grew by 40% in 2009. This was achieved despite the financial crisis that roiled the wind power industry in 2009, and the significant reductions in wholesale electricity prices that began in mid- to late-2008 and have continued to the present. A variety of market drivers allowed year-on-year installation growth to persist in 2009: carryover of projects initially planned for completion in 2008 (but, when the production tax credit was extended through 2012, ultimately came online in 2009); elements of the Recovery Act, including the Section 1603 Treasury Grant Program; the expiration of bonus depreciation rules at the end of 2009; state renewables portfolio standards (RPS); concerns about global climate change; and continued uncertainty about the future costs and liabilities of natural gas and coal facilities.

The yearly boom-and-bust cycle that characterized the U.S. wind power market from 1999 through 2004 – caused by periodic, short-term extensions of the federal production tax credit (PTC) – has now been replaced by five consecutive years of growth. With federal tax incentives for wind energy now extended through 2012, significant capacity additions and a semblance of near-term market stability might be expected. On the other hand, the global financial crisis, lower wholesale electricity prices, and lower demand for renewable energy have created expectations for a slower pace of wind power development in 2010. Moreover, wind power capacity additions in 2009 were buoyed, in part, by projects that were initially slated to be completed in 2008 but that carried over into 2009 when the PTC was extended, somewhat masking the underlying challenges facing the sector. With the extension of federal incentives through 2012, there is less motivation to complete projects in 2010.

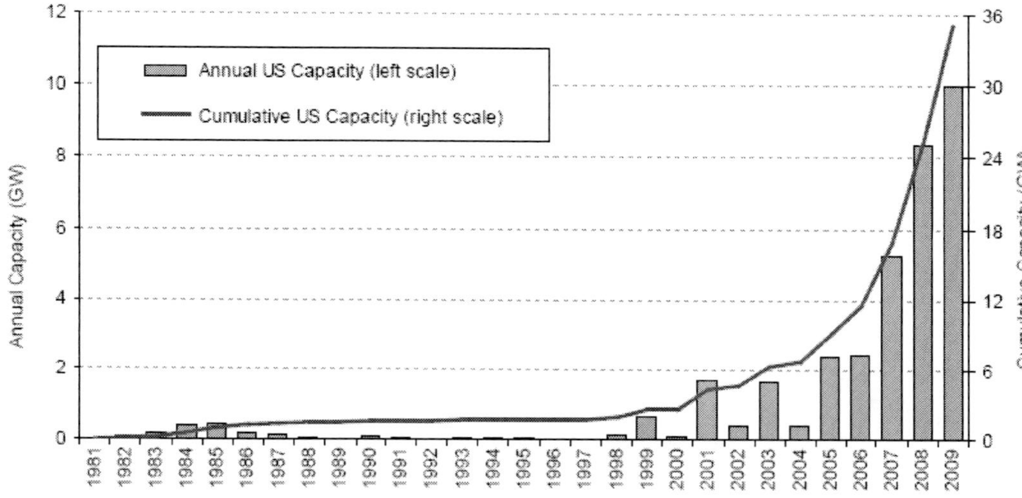

Source: AWEA project database

Figure 1. Annual and Cumulative Growth in U.S. Wind Power Capacity.

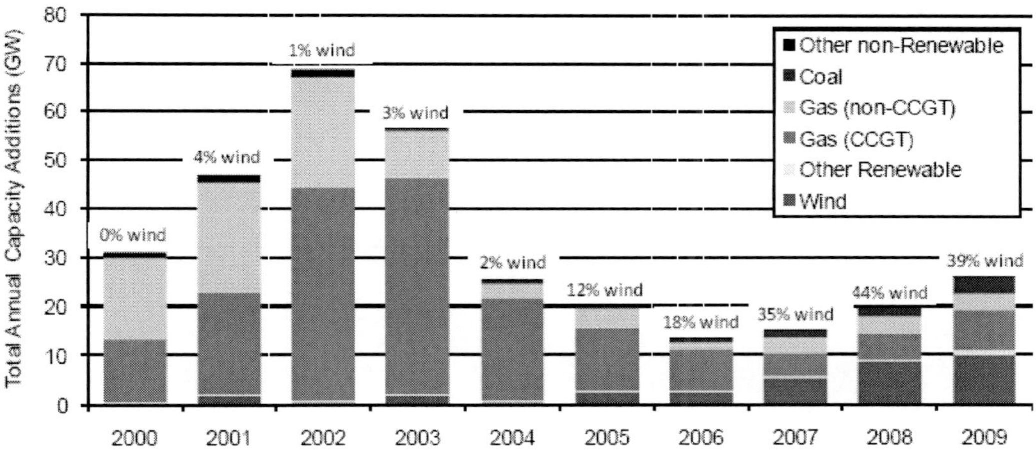

Source: EIA, Ventyx, AWEA, IREC, Berkeley Lab

Figure 2. Relative Contribution of Generation Types in Annual Capacity Additions.

Wind Power Contributed 39% of All New U.S. Electric Generating Capacity in 2009

Wind power now represents one of the largest new sources of electric capacity additions in the United States. For the fifth consecutive year, wind power was the second-largest new resource added to the U.S. electrical grid in terms of aggregate capacity, behind the 11,500 MW of new natural gas plants added in 2009, but ahead of the 3,200 MW of new coal. New wind power projects contributed roughly 39% of the new nameplate capacity added to the U.S. electrical grid in 2009, compared to 44% in 2008, 35% in 2007, 18% in 2006, 12% in 2005, and less than 4% from 2000 through 2004 (see Figure 2).[5]

SMALL WIND TURBINES

Small wind turbines can provide power directly to homes, farms, schools, businesses, and industrial facilities, offsetting the need to purchase some portion of the host's electricity from the grid; such wind turbines can also provide power to off-grid sites. Wind turbines used in these applications are often much smaller – generally ranging in size from a few hundred watts to up to 100 kW or more – than the larger-scale turbines that are the primary focus of this report.

The table below summarizes sales of small wind turbines 100 kW and less in size into the U.S. market. As shown, more than 20 MW of small wind turbines were sold in the U.S. in 2009; most of this new capacity came from turbines manufactured by U.S. companies. These installation figures represent a 15% growth in annual sales – in capacity terms – relative to 2008, yielding a cumulative installed capacity of small wind turbines in the United States in this turbine size range of roughly 100 MW. Within this market segment, there has been a trend towards larger, grid-tied systems (AWEA 2010b).

Year	Annual Sales of Small Wind Turbines into the United States		
	Number of Turbines	Capacity Additions	Sales Revenue
2005	4,324	3.3 MW	$10 million
2006	8,329	8.6 MW	$33 million
2007	9,092	9.7 MW	$42 million
2008	10,386	17.4 MW	$73 million
2009	9,800	20.3 MW	$82 million

Source: AWEA (2010b)

Growth in this sector has been driven – at least in part – by a variety of state incentive programs. In addition, wind turbines equal to or under 100 kW in size are now eligible for an uncapped 30% investment tax credit (the 30% tax credit, with a dollar cap, was initially enacted in October 2008; the cap was removed in the Recovery Act of February 2009).

EIA's (2010) reference-case forecast projects that total U.S. electricity supply will need to increase at an average pace of roughly 49 TWh per year from 2010 to 2035 in order to meet demand growth. On an energy basis, the annual amount of electricity expected to be generated by the new wind power capacity added in 2009 represents nearly 60% of this average annual projected growth in supply.[6] By extension, if wind power additions continued through 2035 at the same pace as set in 2009, then nearly 60% of the nation's projected increase in electricity generation from 2010 through 2035 would be met with wind electricity. Although future growth trends are hard to predict, it is clear that a significant portion of the country's new generation needs is already being met by wind.

The United States Continued to Lead the World in Cumulative Wind Power Capacity, but Was Overtaken by China in Annual Additions

On a worldwide basis, more than 38,000 MW of wind power capacity was added in 2009, the highest volume achieved in a single year, and up from about 28,000 MW in 2008, bringing the cumulative total to approximately 160,000 MW (Table 1). In terms of cumulative installed wind power capacity, the United States ended the year with 22% of total worldwide capacity, and is the leading market in the world by this metric (Table 1 and Figure 3). Over the past 10 years, cumulative wind power capacity has grown an average of 30% per year in the United States, slightly higher than the 28% growth rate in worldwide capacity.

After four years of leading the world in annual wind power capacity additions, the U.S. dropped to second place in 2009 (Table 1), capturing roughly 26% of the worldwide market (behind China's 36% market share[7]), down from 29% in 2008 and 27% in 2007 (Figure 3). Spain, Germany, and India rounded out the top five countries in 2009 for annual capacity additions.[8]

Table 1. International Rankings of Wind Power Capacity

Annual Capacity (2009, MW)		**Cumulative Capacity (end of 2009, MW)**	
China	13,750	**U.S.**	**35,155**
U.S.	**9,994**	China	25,853
Spain	2,331	Germany	25,813
Germany	1,917	Spain	18,784
India	1,172	India	10,827
Italy	1,114	Italy	4,845
France	1,104	France	4,775
U.K.	1,077	U.K.	4,340
Canada	950	Portugal	3,474
Portugal	645	Denmark	3,408
Rest of World	4,121	*Rest of World*	22,806
TOTAL	**38,175**	**TOTAL**	**160,080**

Source: BTM Consult; AWEA project database for U.S. capacity

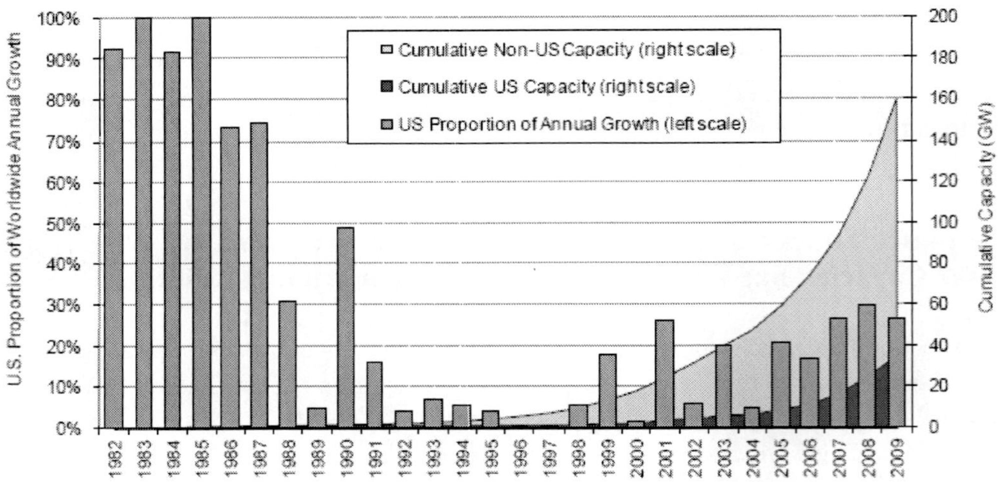

Source: Earth Policy Institute, BTM Consult, AWEA project database

Figure 3. The Contribution of U.S. Wind Power Capacity to Global Wind Power Capacity.

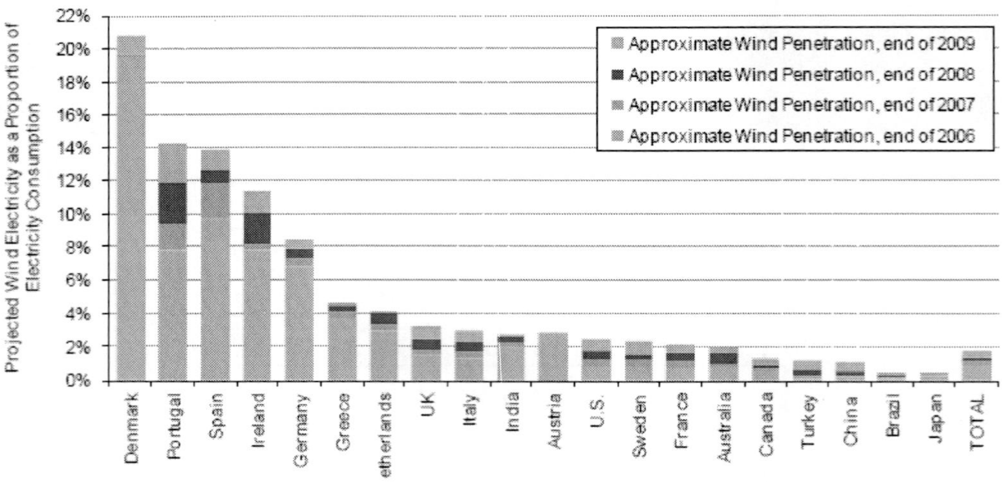

Source: Berkeley Lab estimates based on data from BTM Consult and elsewhere

Figure 4. Approximate Wind Energy Penetration in the Twenty Countries with the Greatest Installed Wind Power Capacity.

Several countries are beginning to achieve relatively high levels of wind energy penetration in their electricity grids. Figure 4 presents data on end-of-2009 (and end-of-2006/07/08) installed wind power capacity, translated into projected annual electricity supply based on assumed country-specific capacity factors, and divided by projected 2010 (and 2007/08/09) electricity consumption. Using this approximation for the contribution of wind power to electricity consumption, and focusing only on the 20 countries with the greatest cumulative installed wind power capacity, end-of-2009 installed wind power is projected to supply the equivalent of roughly 20% of Denmark's electricity demand, 14% of Portugal's, 14% of Spain's, 11% of Ireland's, and 8% of Germany's. In the United States, the cumulative wind power capacity installed at the end of 2009 would, in an average year, be able to supply roughly 2.5% of the nation's electricity consumption (up from 1.8% at the end of 2008, 1.2% at the end of 2007, and 0.8% at the end of 2006).[9] On a global basis, wind energy's contribution at the end of 2009 is estimated to be 1.8%.

Texas Achieved Higher Annual Capacity Additions than Other States, While Four States Have Surpassed 10% Wind Energy Penetration

New large-scale[10] wind turbines were installed in 29 states in 2009. Texas again dominated in terms of new wind power capacity, with 2,292 MW installed in 2009 alone, down somewhat from 2,671 MW installed in 2008. As shown in Figure 5 and Table 2, other leading states in terms of new capacity (each with more than 500 MW) include Indiana, Iowa, Oregon, Illinois, New York, and Washington. Thirteen states added more than 200 MW each in 2009.

On a cumulative basis, Texas continued to build on its lead in 2009, with a total of 9,410 MW of wind power capacity installed by the end of the year. In fact, Texas has more installed wind power capacity than all but five countries worldwide. U.S. states following Texas in cumulative installed capacity are Iowa, California, Washington, and Oregon. Sixteen states had more than 500 MW of wind power capacity as of the end of 2009, with fourteen topping 1,000 MW, and three topping 2,000 MW. Although all wind power projects in the United States to date have been installed on land, offshore development activities continued in 2009, as discussed in the next section.

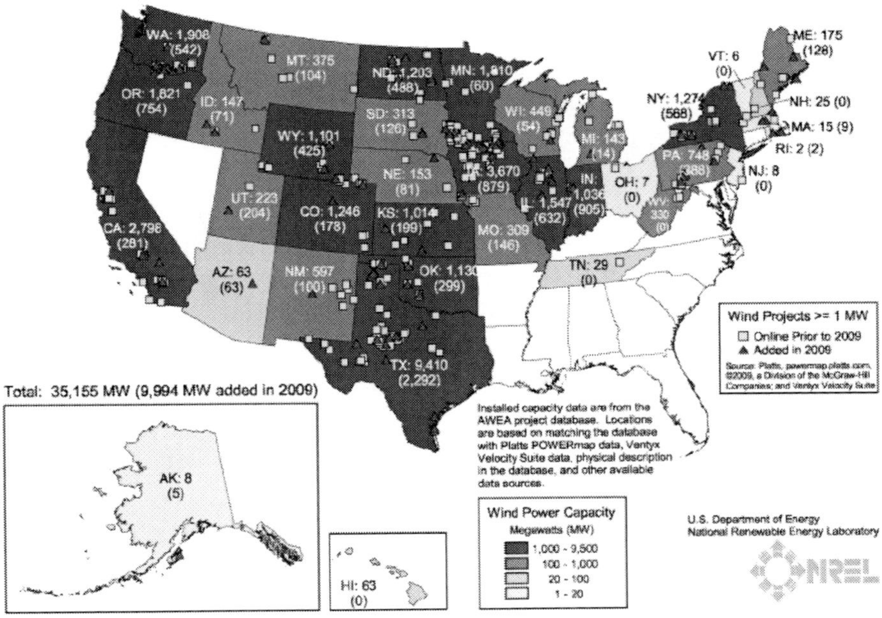

Note: Numbers within states represent cumulative installed wind capacity and, in parentheses, annual additions in 2009.

Figure 5. Location of Wind Power Development in the United States.

Table 2. United States Wind Power Rankings: The Top 20 States

Annual Capacity (2009, MW)		Cumulative Capacity (end of 2009, MW)		*Estimated* Percentage of In-State Generation*	
Texas	2,292	Texas	9,410	Iowa	19.7%
Indiana	905	Iowa	3,670	South Dakota	13.3%
Iowa	879	California	2,798	North Dakota	11.9%
Oregon	754	Washington	1,908	Minnesota	10.7%
Illinois	632	Oregon	1,821	Oregon	9.0%
New York	568	Minnesota	1,810	Colorado	7.7%
Washington	542	Illinois	1,547	Kansas	7.4%
North Dakota	488	New York	1,274	Texas	6.8%
Wyoming	425	Colorado	1,246	Wyoming	6.7%
Pennsylvania	388	North Dakota	1,203	Oklahoma	5.0%
Oklahoma	299	Oklahoma	1,130	Montana	4.9%
California	281	Wyoming	1,101	Washington	4.9%
Utah	204	Indiana	1,036	New Mexico	4.6%
Kansas	199	Kansas	1,014	California	3.4%
Colorado	178	Pennsylvania	748	Maine	3.1%
Missouri	146	New Mexico	597	Idaho	3.0%
Maine	128	Wisconsin	449	Indiana	2.7%
South Dakota	126	Montana	375	New York	2.2%
Montana	104	West Virginia	330	Hawaii	2.2%
New Mexico	100	South Dakota	313	Illinois	2.1%
Rest of U.S.	358	*Rest of U.S.*	1,376	*Rest of U.S.*	0.3%
TOTAL	**9,994**	**TOTAL**	**35,155**	**TOTAL**	**2.5%**

* Based on a projection of wind electricity generation from end-of-2009 wind power capacity, divided by total instate electricity generation in 2009.

Source: AWEA project database, EIA, Berkeley Lab estimates

Some states are beginning to realize relatively high levels of wind energy penetration. Table 2 lists the top 20 states based on an estimate of wind electricity generation from end-of-2009 wind power capacity, divided by total in-state electricity generation in 2009.[11] By this metric, four Great Plains states lead the list in terms of estimated wind energy as a percentage of total in-state generation. Specifically, the wind power capacity installed as of the end of 2009 is estimated, in an average year, to be capable of generating approximately 19.7% of all in-state electricity generation in Iowa, 13.3% in South Dakota, 11.9% in North Dakota, and 10.7% in Minnesota. Five additional states surpass the 5% mark by this metric, while twenty states exceed 2%.

Some utilities are achieving even higher levels of wind energy penetration into their individual electric systems. Table 3 lists the top-20 utilities in terms of aggregate wind power capacity on their systems at the end of 2009, based on data provided by AWEA (2010a). Included here are wind power projects either owned by or under long-term contract with these utilities for use by their own customers; short-term renewable electricity and renewable energy certificate purchases are excluded. The table also lists the top-20 utilities based on an estimate of the percentage of retail sales that wind electricity represents, using end-of-2009 wind power capacity, wind power capacity factors that are consistent with the state or region in which a utility operates, and EIAprovided aggregate retail electricity sales for each utility in 2008.[12] As shown, Minnkota Power Cooperative is estimated to have achieved 38% wind energy penetration by this metric, while a total of nine utilities are estimated to have exceeded 10%.

Offshore Wind Power Project and Policy Developments Accelerated in 2009

Offshore wind power projects totaling 689 MW were installed globally in 2009, bringing worldwide offshore capacity to 2,110 MW (BTM 2010). The vast majority of this capacity is located in Europe. In contrast, all wind power projects built in the United States to date have been sited on land. The availability of low-cost land-based wind energy, regulatory delays and uncertainty associated with offshore development, turbine supply shortages, high and uncertain offshore project costs, and public acceptance concerns have so far hampered progress in the offshore sector in the United States. Nonetheless, there is interest in offshore wind energy in several parts of the country, driven by the proximity of offshore wind resources to large population centers, advances in technology, potential local economic development benefits, and superior capacity factors (and, in some instances, peak load coincidence) compared to the finite set of attractive and developable land-based wind power projects available in some regions.

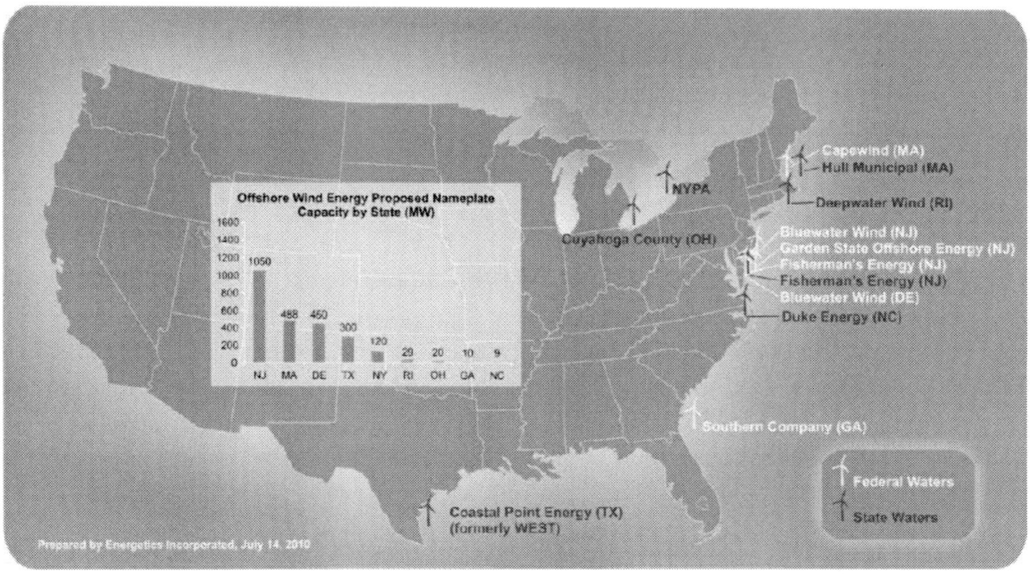

Source: Musial et al. (2010)

Figure 6. Offshore Wind Power Projects in a Relatively Advanced State of Development.

Table 3. Top-20 Utility Wind Power Rankings

Total Wind Capacity (end of 2009, MW)		*Estimated* Percentage of Retail Sales (for utilities with > 100 MW of wind)*	
Xcel Energy	3,176	Minnkota Power Cooperative	38.0%
MidAmerican Energy	2,923	Empire District Electric Company	18.1%
Southern California Edison	1,772	Turlock Irrigation District	18.0%
American Electric Power	1,196	Otter Tail Power	14.0%
Pacific Gas & Electric	1,131	Sunflower Electric Power Corp.	13.2%
Luminant	913	Xcel Energy	11.1%
Alliant Energy	645	Austin Energy	10.3%
City Public Service of San Antonio	579	Great River Energy	10.1%
Puget Sound Energy	479	Westar	10.1%
Austin Energy	439	Western Farmers' Electric Cooperative	9.8%
First Energy	376	MidAmerican Energy	9.6%
Portland General Electric	375	Snohomish Public Utility District	8.5%
Minnkota Power Cooperative	357	MSR Public Power Agency	8.4%
Basin Electric	352	City Public Service of San Antonio	8.4%
SDG&E	342	Public Service New Mexico	6.8%
Great River Energy	319	Cowlitz Public Utility District	6.5%
Westar	295	WPPI Energy	6.4%
Oklahoma Gas & Electric	272	Alliant Energy	5.9%
Empire District Electric Company	255	Puget Sound Energy	5.4%
SCPPA (not including LADWP)	233	Northwestern Energy	5.3%

* Based on a projection of wind electricity generation from end-of-2009 wind power capacity, divided by the retail sales for each utility in 2008.

Source: AWEA, EIA, Berkeley Lab estimates

Figure 6 shows 13 proposed offshore wind power projects in the United States that have advanced significantly in the permitting and development process. These projects have either made substantial progress towards receiving state approval or have received a lease or "interim limited lease" from the U.S. Minerals Management Service (MMS).[13] In total, these proposed projects equal 2,476 MW, and are primarily located in the Northeast and Mid-Atlantic, though notable proposed projects also exist in the Southeast, Great Lakes, and Gulf of Mexico. Even these "advanced stage" projects are in various stages of development – some or even many may never be realized, while other projects not identified in the figure are also under consideration.

Of the projects identified in Figure 6, three have signed or proposed power purchase agreements with terms and details that have been made public (see Table 4). As shown, offshore wind energy prices are substantially greater than those presented later for land-based wind energy.

Table 4. Proposed Power Purchase Agreements for Offshore Wind Power Projects.

Seller	Purchaser	Location / Amount	Contract Details
NRG Bluewater	Delmarva	Delaware 200 MW*	25-yr contract for electricity and a portion (28.6%) of the RECs: $132/MWh in 2013, escalating at 2.5%/yr; approved by regulatory commission in July 2008
Deepwater Wind	National Grid	Rhode Island 28.8 MW	20-yr contract for electricity and RECs: $244/MWh in 2013, escalating at 3.5%/yr; filed with regulatory commission in June 2010
Cape Wind	National Grid	Massachusetts 50% of 468 MW**	15-yr contract for electricity and RECs: $187/MWh in 2013, escalating at 3.5%/yr; filed with regulatory commission in May 2010 with revisions in July 2010

Source: Berkeley Lab review of regulatory filings

* NRG Bluewater has contracted for an additional ~93 MW from their 450 MW proposed Delaware facility under a memorandum of understanding with the Delaware Electric Municipal Corporation and a contract with the University of Maryland.

** National Grid is also seeking approval of a second nearly-identical but conditional PPA for the remaining 50% that is intended to be available for assignment to other parties in the future and is intended to facilitate project financing.

Several other project-level announcements in 2009 and early 2010 demonstrate the accelerating pace of offshore wind power development in the United States. Most notably, after nine years in the permitting process, the 468 MW Cape Wind project was granted approval by MMS in April 2010, following the 2009 completion of MMS's Final Environmental Impact Statement as well as the state and local permitting process. Cape Wind also selected a turbine supplier in 2010 (Siemens 3.6 MW turbines), received FAA approval, and filed a PPA for consideration before the state's utility regulatory commission.[14]

In Delaware, NRG Bluewater Wind was awarded an interim limited lease by MMS in 2009, and a contract with the University of Maryland was announced for an additional 55 MW of the project's output (i.e., in addition to the 200 MW contract with Delmarva shown in Table 4). In New Jersey, NRG Bluewater Wind, Garden State Offshore Energy, and Fisherman's Energy were all awarded interim limited leases by MMS in 2009, and each is benefiting from a state funding cost share for meteorological testing. Garden State Offshore Energy, a joint venture between Public Service Enterprise Group and Deepwater Wind, was selected by the New Jersey Board of Public Utilities in 2008 to receive a $4 million grant from the state in support of a 350 MW offshore project, with additional agreements with NRG Bluewater Wind and Fisherman's Energy following. In Rhode Island, the PPA for the 28.8 MW Block Island demonstration project was initially rejected by the state's public utilities commission in March 2010, but subsequent state legislation has led to contract modifications and re-submittal to the commission in June 2010 for further consideration; in early 2009, Rhode Island signed a joint development agreement with Deepwater Wind under which Deepwater was selected as the state's preferred offshore wind power project developer for both the demonstration project and a subsequent, larger project. In Ohio, the final feasibility study for a 20 MW offshore wind power project was completed in 2009, and an agreement was signed in 2010 for the purchase of five 4 MW GE turbines; GE's purchase in 2009 of offshore turbine manufacturer ScanWind presumably helped enable the agreement. In December 2009, the New York Power Authority issued a request for proposals to select developers for projects of at least 120 MW (and up to 500 MW) in Lake Erie and/or Lake Ontario; five responses were received, and selections are expected in late 2010 or early 2011. The Long Island Power Authority, Consolidated Edison, and the New York Power Authority are also collaborating on a possible offshore wind power project off the coast on Long Island, with an application for a federal lease currently anticipated in 2010. In North Carolina, Duke Energy signed a contract in October 2009 with UNC Chapel Hill to install three demonstration turbines in Pamlico Sound; Duke Energy will pay for the turbines and their installation. Finally, Texas has state authority over permitting to 10 miles offshore, and a number of leases have been issued to projects in various stages of development.

In addition to these project-level developments, policy and regulatory activity related to offshore wind energy deployment continued. Following the early 2009 announcement of MMS's rules governing offshore wind power development leases, easements, and royalties, the federal government has moved towards implementation. The creation of a new regional MMS office to coordinate and appropriately expedite renewable energy development on the Atlantic Outer Continental Shelf was announced in 2010, for example, as was the creation of the Atlantic Offshore Wind Energy Consortium, through which the Department of the Interior and East Coast states will facilitate federal-state cooperation on offshore wind power development on the Outer Continental Shelf. Earlier in 2009, MMS and FERC came to agreement on their respective roles in offshore energy development. MMS subsequently began to issue limited leases for five years of resource testing under an interim policy. Then, in April 2010, MMS released a Request for Information (RFI) for commercial leasing for wind power on the outer continental shelf off of Delaware. The RFI invites interested parties to submit descriptions of their interest in obtaining a commercial lease in specific areas off the coast of Delaware. The RFI details information that developers should submit and all parties are invited to submit information on environmental issues of concern. MMS (now called the Bureau of Ocean Energy) is required to issue leases competitively, and the RFI is the first step in determining if there is competitive interest in wind energy off the Delaware coast.

At the state level, the final Massachusetts Ocean Management Plan was released in December 2009, which (among other provisions) encourages community-scale offshore wind power development, creates a formal role for regional planning authorities in offshore energy planning, and identifies two larger designated offshore wind energy areas in state waters. In Maine, legislation was passed in 2010 establishing a goal for at least 300 MW of offshore wind energy by 2020 and 5,000 MW by 2030, and implementing other regulatory changes intended to facilitate offshore wind power development in the state; Maine also received Recovery Act funds from DOE to support offshore wind energy research, testing, and demonstration projects. Finally, in late 2009, three Mid-Atlantic states (Maryland, Delaware, and Virginia) signed an agreement to work together collaboratively on regional offshore wind power development.

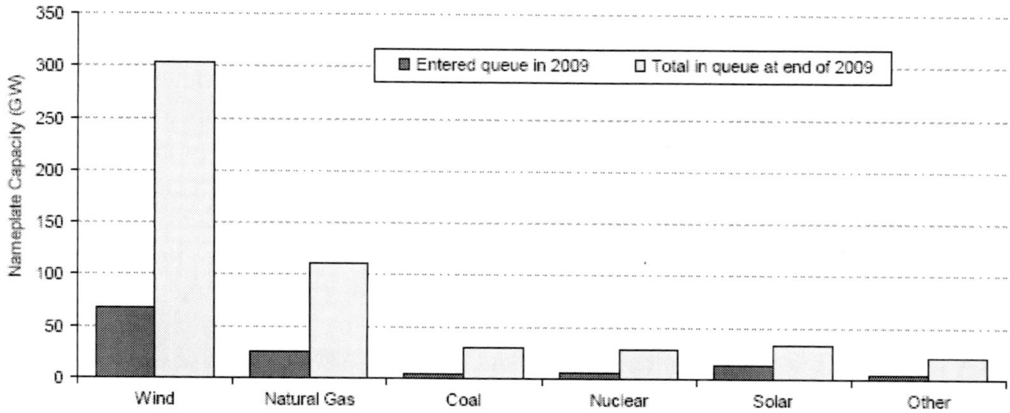

Source: Exeter Associates review of interconnection queues

Figure 7. Nameplate Resource Capacity in 33 Selected Interconnection Queues.

Data from Interconnection Queues Demonstrate that an Enormous Amount of Wind Power Capacity Is Under Consideration

One visible testament to the increased interest in wind energy is the amount of wind power capacity currently working its way through the major transmission interconnection queues across the country. Figure 7 provides this information for wind power and other resources aggregated across 33 different interconnection queues administered by independent system operators (ISOs), regional transmission organizations (RTOs), and utilities.[15] These data should be interpreted with caution: though placing a project in the interconnection queue is a necessary step in project development, being in the queue does not guarantee that a project will actually get built. In fact, there is a growing recognition that many of the projects currently in interconnection queues are very early in the development process. As a result, efforts have been and are being taken by the Federal Energy Regulatory Commission (FERC), ISOs, RTOs, and utilities to reduce the number of speculative projects that have – in recent years – clogged these queues (Porter et al. 2009).

Even with this important caveat, the amount of capacity in the nation's interconnection queues still provides at least some indication of the amount of wind power development that is in the planning phase. At the end of 2009, even after reforms by a number of ISOs, RTOs, and utilities to reduce the number of projects in their queues, there were roughly 300 GW of wind power capacity within the interconnection queues reviewed for this report – nearly nine times the installed wind power capacity in the United States.[16] This wind power capacity represented almost *60%* of all generating capacity within these selected queues at that time, and was nearly *three times* as much capacity as the next-largest resource in these queues, natural gas.

Much of this wind power capacity is planned for the Midwest, Mountain, Texas, PJM, SPP, and Northwest regions: wind power projects in the interconnection queues in these regions account for 93% of the aggregate 303 GW of wind power in the selected queues (see Figure 8). At the other end of the spectrum, smaller amounts of wind power capacity are represented in the interconnection queues of the California ISO (3.4%), New York ISO (2.3%), ISO-New England (1.2%), and the Southeast (0.1%).

As another data point, the North American Electric Reliability Corporation (NERC) finds that roughly 210 GW of new wind power capacity is planned for construction over the next ten years in the United States (NERC 2009a). Once again, though, it is unlikely that all of these planned projects will ultimately come to fruition within this time frame. As a measure of the near-term development pipeline, Ventyx (2010) estimates that – as of late-July 2010 – more than 31 GW of wind power capacity was either under construction or in site preparation (5 GW of the 31 GW total), in-development and permitted (12 GW of the 31 GW), or in-development with pending permit and/or regulatory applications (the remaining 14 GW of the 31 GW total). AWEA (2010c), meanwhile, identified 5,700 MW of wind power projects that were under construction as of mid-2010.

3. Industry Trends

GE Remained the Top Turbine Manufacturer in the U.S. Market, but a Growing Number of Other Manufacturers Are Capturing Market Share

General Electric (GE) remained the number one manufacturer of wind turbines supplying the U.S. market in 2009, with 40% of domestic turbine installations (down slightly from 43% in 2008, 45% in 2007, and 47% in 2006).[17] Following GE were Vestas (15%), Siemens (12%), Mitsubishi (8%), Suzlon (7%), Clipper (6%), Gamesa (6%), REpower (3%), Acciona (2%), and Nordex (1%). Other utility-scale (>100 kW) wind turbines installed in the United States in 2009 (and that fall into the "Other" category in Figure 9) include turbines from NedWind (6.5 MW), AAER (6 MW), DeWind (6 MW), Fuhrlander (4.5 MW), Goldwind (4.5 MW), RRB (2.4 MW), Elecon (0.6 MW), and Wind Energy Solutions (0.25 MW).

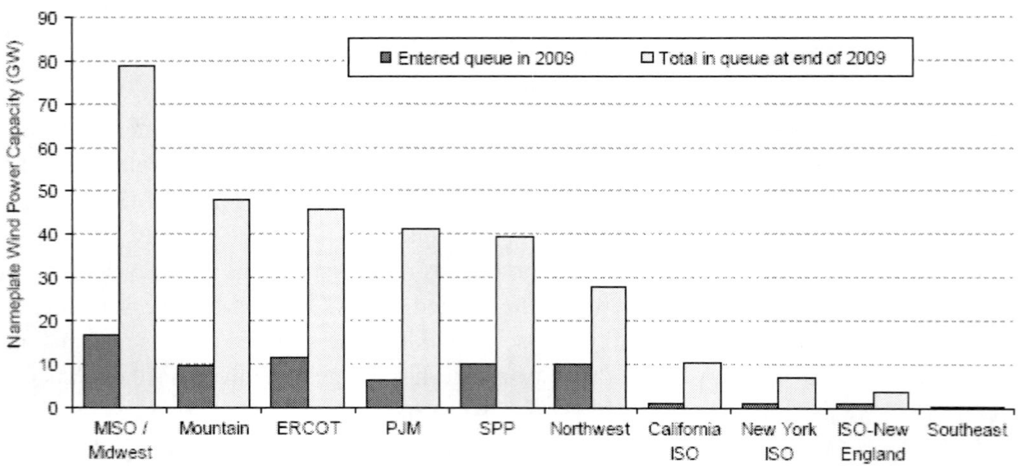

Source: Exeter Associates review of interconnection queues

Figure 8. Wind Power Capacity in 33 Selected Interconnection Queues.

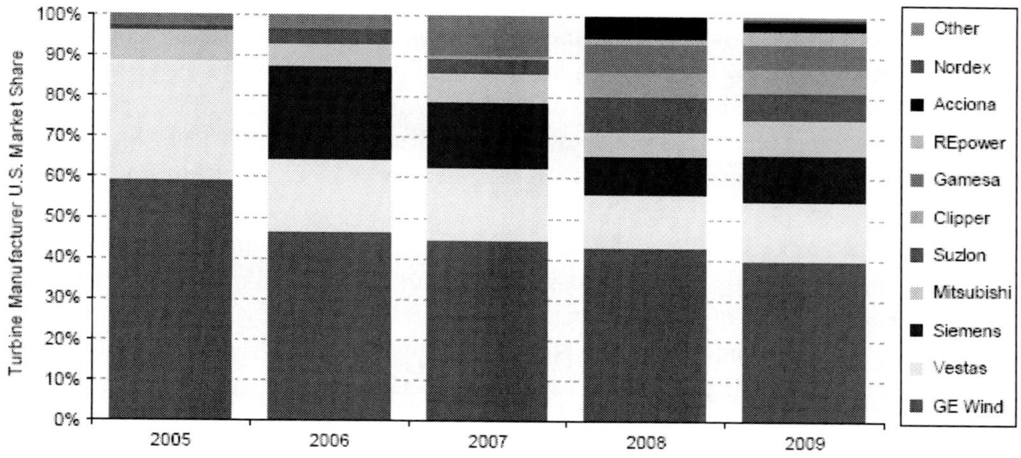

Source: AWEA project database

Figure 9. Annual U.S. Market Share of Wind Manufacturers by MW, 2005-2009.

A notable increase in competition among wind turbine manufacturers has occurred since 2005, with the number of manufacturers installing more than 1 MW increasing from just 6 in 2005 to 16 manufacturers in 2009. Consequently, the market share of the leading manufacturers – in percentage terms – has generally declined. Manufacturers with modern wind turbines installed in the United States now hail from not just the United States, Europe, and Japan, but also from India and, for the first time in 2009, China. Chinese and South Korean manufacturers, in particular, began to express strong interest in entering the U.S. market in 2009, though the timing and speed of that entry remains uncertain.

Table 5. Annual U.S. Turbine Installation Capacity, by Manufacturer

Manufacturer	Turbine Installations (MW)				
	2005	**2006**	**2007**	**2008**	**2009**
GE Wind	1,433	1,146	2,342	3,585	3,995
Vestas	700	439	948	1,120	1,490
Siemens	0	573	863	791	1,162
Mitsubishi	190	128	356	516	814
Suzlon	25	92	197	736	702
Clipper	3	0	48	470	605
Gamesa	50	74	494	616	600
REPower	0	0	0	94	330
Acciona	0	0	0	410	204
Nordex	0	0	3	0	63
Other	2	2	0	12	31
TOTAL	**2,402**	**2,454**	**5,249**	**8,350**	**9,994**

Source: AWEA project database

Notwithstanding any changes in market share that have occurred in percentage terms, most manufacturers saw installations of their turbines grow between 2008 and 2009, in many cases significantly. As shown in Table 5, the most significant growth was experienced by GE (+410 MW), Vestas and Siemens (+370 MW each), Mitsubishi (+298 MW), and REpower (+236 MW). Suzlon and Gamesa each installed a few less MW in the United States in 2009 than in 2008, while Acciona's installations declined by a more significant amount (-206 MW).

In 2009, U.S.-owned GE was the second-leading supplier of turbines *globally*, with a 12.4% market share, slightly behind Vestas' 12.5% market share. Clipper was the 13th largest manufacturer, with 1.6% of the worldwide market (BTM 2010).[18] On a worldwide basis, perhaps the most significant story of 2009 was the growing market share of Chinese turbine manufactures; to date, that growth has been based almost entirely on sales to the Chinese market.

Domestic Wind Turbine and Component Manufacturing Investments Remained Strong in 2009, but the Financial Crisis and Weak Turbine Sales Slowed the Sector's Growth

As wind power deployment has increased in the United States, a growing number of foreign and domestic turbine and component manufacturers have begun or continued to localize and expand operations across the nation.

Though the financial crisis resulted in a slowdown in overall U.S. manufacturing in 2009, wind equipment manufacturing was, to a degree, a bright spot. Figure 10 presents a non-exhaustive list of 13 wind turbine and component manufacturing and assembly facilities that opened in 2009, and identifies their location. The map also depicts the location of 21 new manufacturing facilities announced (but not yet built) in 2009, identifies 12 existing firms that expanded into the wind energy sector in 2009, and plots 125 existing turbine and component manufacturing facilities that opened prior to 2009.

Of the 34 new or announced facilities in 2009 captured in Figure 10, two are owned by major international wind turbine original equipment manufacturers (OEMs): Siemens (nacelles in Hutchinson, Kansas), and Mitsubishi Power Systems (nacelles in Fort Smith, Arkansas). In addition, GE announced plans to open a research facility in Van Buren Township, Michigan. (Research facilities are not included in the figure).

Several smaller- to mid-sized OEMs also opened or announced U.S. factories in 2009. Nordic Windpower, for example, opened a turbine manufacturing and assembly facility in Pocatello, Idaho, where it is producing and assembling its 1 MW turbines. Continental Wind announced that it would open a turbine manufacturing facility in Santa Paula, California, where it will reportedly focus on the production of turbines ranging from 300-900 kW. Still other firms, including a number of major international turbine vendors, continued to make progress in establishing themselves in the U.S. market based on manufacturing plans announced prior to 2009.

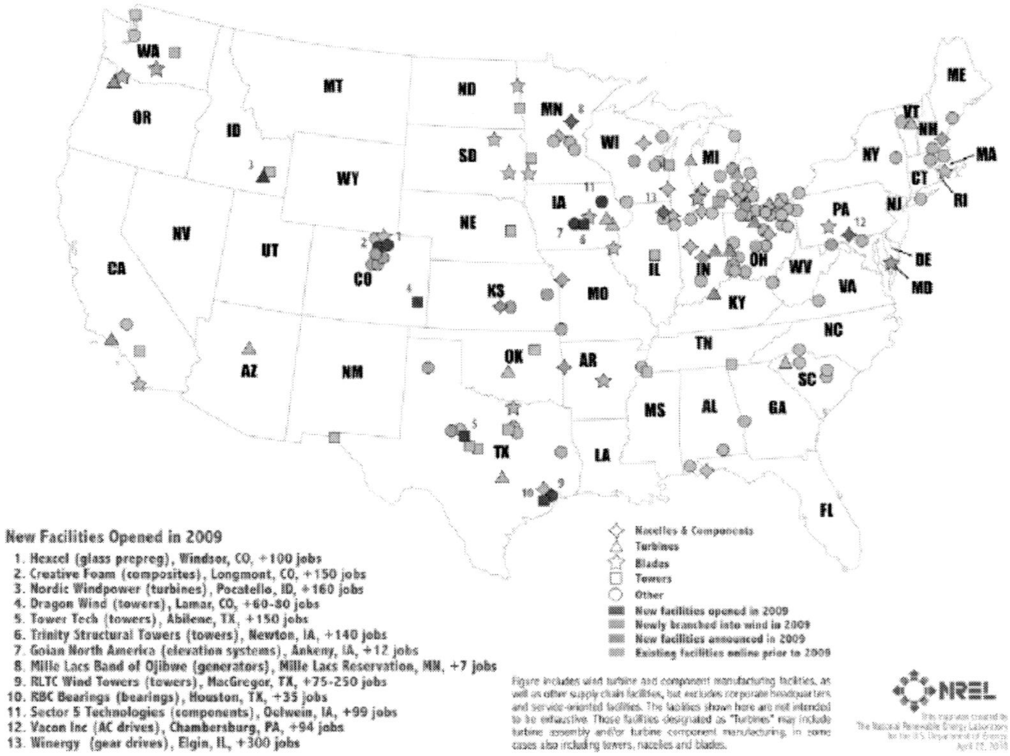

Figure 10. Location of Existing and New Turbine and Component Manufacturing Facilities.

As a result of this activity, seven of the ten OEMs with the largest share of the U.S. market in 2009 (GE, Vestas, Siemens, Suzlon, Clipper, Gamesa, Acciona) have one or more manufacturing facilities operating in the United States, and two of the remaining three (Mitsubishi, Nordex) have announced specific plans to open facilities in the future. These figures compare to just one utility-scale wind turbine OEM assembling nacelles in the United States in 2004 (GE). Still other active domestic and foreign OEMs have already established manufacturing facilities in the United States (Nordic, DeWind) or have at least tentatively announced the location of future U.S. manufacturing facilities (Alstom, Emergya, Fuhrlander, A-Power, Ming Yang),[19] while several U.S. companies have announced their interest in manufacturing but have not yet built any utility- scale turbines (e.g., Northern Power Systems, Continental Wind).

Other notable developments from 2009 and early 2010 include the cash infusions garnered by emerging domestic turbine OEMs, and the increased interest in the United States market by Asian players. Clipper, for example, received a cash infusion through the sale of 49.5% of the company to United Technologies Corporation, while Nordic Windpower also raised substantial new capital. Meanwhile, in addition to Mitsubishi's planned nacelle assembly facility, new Asian OEMs from South Korea and China demonstrated interest in the U.S. market. South Korea's Samsung, Hyundai, and Unison, for example, all announced interest in U.S. sales of wind turbines, while California-based Composite Technology Corporation sold its DeWind manufacturing business to South Korea's Daewoo Shipbuilding & Marine Engineering. China's Goldwind installed its first wind turbines on United States soil in 2009, and a number of other Chinese manufacturers have also announced their entry into the market; in early 2010, for example, A-Power and Ming Yang announced the location of possible future manufacturing facilities in Nevada and Texas, respectively.

Figure 10 shows a considerable number of new component manufacturing facilities announced or opened in 2009, from both foreign and domestic firms. A number of domestic manufacturers that were not previously active in the wind energy sector transitioned into the industry in 2009: five such companies are located in Michigan, one of the states hardest hit by the economic downturn. These facilities span the entire supply chain. Though new and announced component manufacturing facilities are spread across the country, a number of companies are choosing to locate near already-established large-scale OEMs; for example, in 2009, six component suppliers announced or opened facilities in Colorado, where Vestas has made significant investments.

Some of the states that have experienced the greatest growth in installed wind power capacity in recent years are also seeing significant new manufacturing activity. Even states with little installed wind power capacity, however, are reaping job and economic benefits from new wind- related manufacturing facilities, particularly if those states are strategically positioned geographically near the main wind power markets and in locations that minimize transportation logistics challenges and costs (e.g., Arkansas).

As a result of these developments, AWEA (2010a) estimates that the wind energy sector employed 85,000 full-time workers in the United States at the end of 2009, the same figure as in 2008 but much higher than in years prior to 2008. In addition to manufacturing, these jobs span project development, construction and turbine installation, operations and maintenance, transportation and logistics, and financial, legal, and consulting services.

Notwithstanding the generally positive outlook for the turbine manufacturing sector in the United States, however, the industry is facing economic headwinds. AWEA (2010a) estimates that 1,500 manufacturing jobs were lost in 2009 (reducing the number of wind turbine and component manufacturing jobs in the United States to around 18,500 at year end, but still up significantly from years past) as a number of firms delayed or scaled-back their expansion plans and announced layoffs as a result of weak demand for wind turbines and the poor state of the U.S. economy. With financial conditions showing signs of stabilizing, some manufacturers have already begun the process of rehiring workers and resuming their expansion plans, but the outlook for 2010 remains uncertain. In addition, as the domestic industry expands, a new challenge has become more acute: workforce training and development for all segments of the wind power industry. A variety of programs at the local and national levels are beginning to target these needs.

A Growing Percentage of the Equipment Used in U.S. Wind Power Projects Has Been Sourced Domestically in Recent Years

As a result of the foregoing developments in U.S.-based wind turbine and component manufacturing, AWEA estimates that the share of *domestically manufactured* wind turbines and components grew from 20-25% in 2005 to roughly 50% in 2009 (AWEA 2010a). David (2010) uses somewhat different methods and assumptions, and estimates that the *import share* of wind turbines and selected components as a proportion of total turbine costs declined from

64% in 2006 to 32% in 2009 (these data suggest a domestic content of 36% in 2006, increasing to 68% in 2009; one reason for the difference with AWEA is that David (2010) focuses on imports as a fraction of *total turbine costs* – including soft costs associated with turbine sales transactions – whereas AWEA (2010a) focuses on imports as a fraction of *turbine equipment costs* alone).[20]

These general trends are confirmed in this section by relying upon similar import data as used by David (2010), specifically data from the U.S. Department of Commerce,[21] but with somewhat different assumptions and data processing. This analysis supports the basic conclusion that the United States remains a large importer of wind power equipment, but that wind power capacity growth is outpacing growth in imports, yielding a growing share of domestic manufacturing content.

Figure 11 presents calendar-year data on U.S. imports and exports of wind-powered generating sets from 2006 through 2009. [22] Wind-powered generating sets include nacelles and, when imported with the nacelle, certain turbine components. (Data on the *separate* importation of turbine components are additive to the data shown in Figure 11, are reported in Figure 12, and are discussed later in this section).

As shown in Figure 11, U.S. imports of wind-powered generating sets grew from $1.3 billion in 2006 to nearly $2.5 billion in both 2007 and 2008, before falling to roughly $2.3 billion in 2009 (all data are presented in real 2009 dollars). At $2.3 billion, the United States was – by far – the largest importer of wind-powered generating sets in 2009, representing approximately 34% of worldwide imports (no other country reached 10% of global imports).[23] The primary source markets from which these imports to the United States originate have been and continue to be the home countries of the major international wind turbine manufacturers: Denmark, Spain, Japan, India, and Germany.

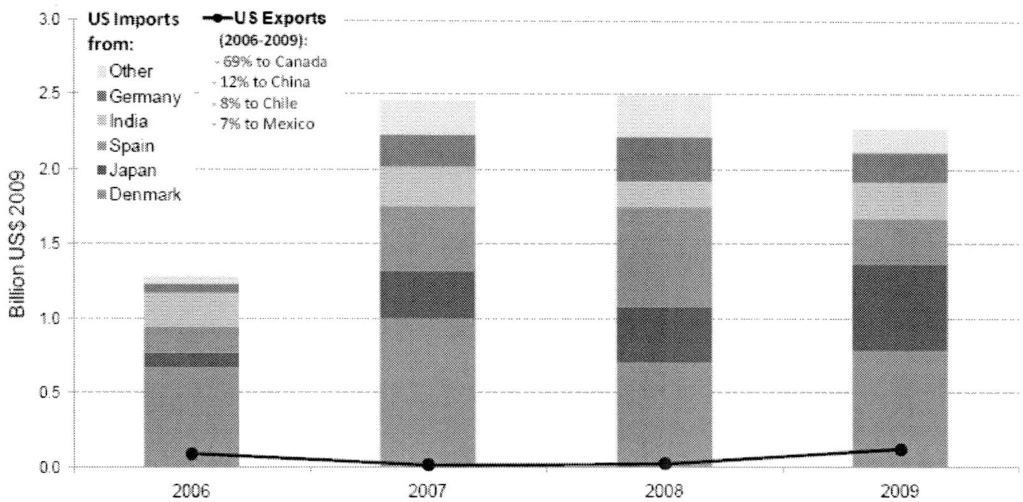

Source: Berkeley Lab analysis of data from USITC Data Web: http://dataweb.usitc.gov

Figure 11. Imports and Exports of Wind-Powered Generating Sets.

Exports of wind-powered generating sets from the United States increased to $120 million in 2009, up from roughly $20 million in 2008. The largest destination markets for U.S. exports over the entire 2006-2009 timeframe included Canada (69%), China (12%), Chile (8%), and Mexico (7%). It is clear from Figure 11 that the United States remains a sizable net importer of wind-powered generating sets. In fact, while the United States ranks first globally as an importer of wind-powered generating sets, representing 34% of global imports, it ranks seventh in the export of similar equipment, representing 3% of global exports, behind the largest *global* exporters of Denmark (28% of exports), Germany (23%), Spain (19%), Japan (12%), and India (8%).

The data presented in Figure 11 are for wind-powered generating sets. Wind turbine blades, hubs, generators, gearboxes, and other components are included in Figure 11 only if shipped with the nacelle itself. These same wind turbine components may also be imported separate from the nacelle, however, implying that the data presented in Figure 11 include only a fraction of total wind equipment imported into the United States. Data for the *separate* importation of some wind turbine components are also available and can be added to the imports shown in Figure 11, but data on the separate importation of turbine components are embedded within larger trade categories that include sectors other than wind energy.

Figure 12 presents *estimated* calendar-year data for the separate importation of selected wind turbine components that include towers (trade category is "towers and lattice masts"), generators ("AC generators from 750 to 10,000 kVA"), blades and other components ("parts of other engines and motors" and "parts of generators"), and gearboxes ("other fixed ratio speed changers" and "other multiple and variable ratio speed changers").[24] The import estimates shown in Figure 12 should be viewed with caution because the underlying data used to produce the figure are based on trade categories that are not exclusive to wind energy (e.g., they could include generators for non-wind applications). The wind turbine component-level import estimates shown in Figure 12 therefore required assumptions about the fraction of larger trade categories likely to be represented by wind turbine components.[25]

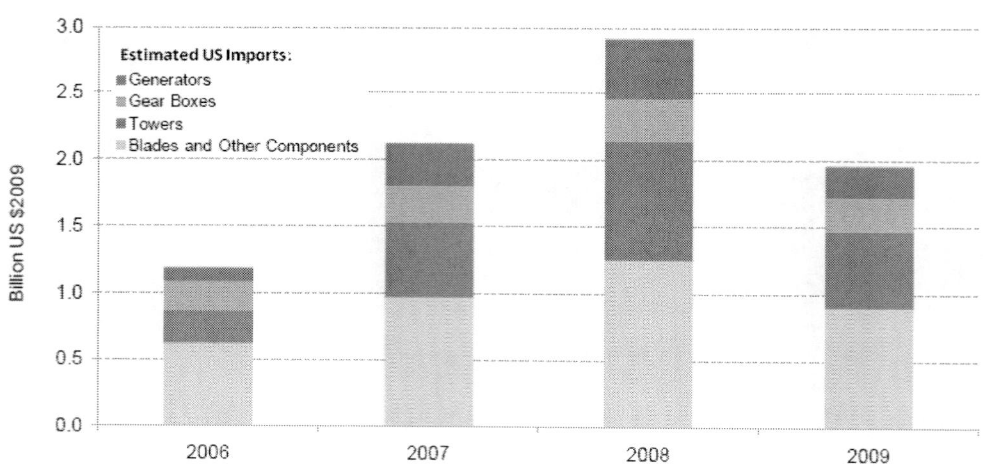

Source: Berkeley Lab analysis of data from USITC Data Web: http://dataweb.usitc.gov

Figure 12. Estimated Imports of Goods in Trade Categories That Include Wind Turbine Components Using Wind-Related Import Percentage Assumptions.

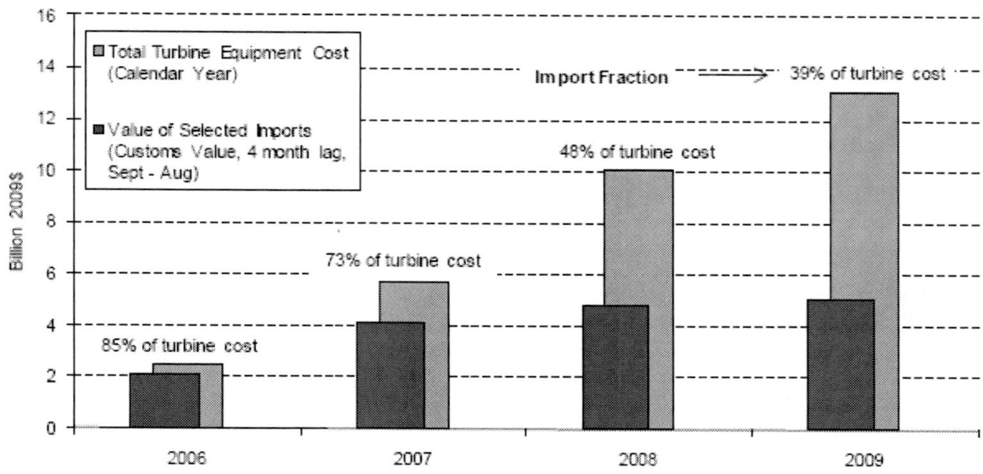

Figure 13. Wind Power Equipment Imports as a Fraction of Total Turbine Cost.

Figure 12 confirms that trends in the *separate* importation of certain wind turbine components is consistent with the import trends for wind-powered generating sets presented in Figure 11. The estimated imports of separate wind turbine components on a calendar-year basis, as presented in Figure 12, increased from $1.2 billion in 2006 to $2.9 billion in 2008, before falling to roughly $2 billion in 2009 (again, all dollar values are expressed in real 2009 dollars). Wind turbine component exports in these trade categories are not shown in the figure because such exports are likely a small (and uncertain) fraction of the broader trade category totals.

Looking behind the data presented in Figure 12 in more regional detail, notable trends include an increase in imports of blades and other components ("parts of other engines and motors" and "parts of generators") from Mexico and India in 2009, and a sizable shift of tower and lattice mast imports away from Europe and Canada and toward Asia. Similarly, from 2006-2009, an increasing share of generator imports have come from Asia, whereas European imports have declined.[26]

Though Figures 11 and 12 depict a U.S. market that remains reliant on imports of wind power equipment, that reliance has declined over time as growth in installed wind power capacity has outpaced growth in wind turbine and component imports. Specifically, adding the data in Figure 11 and Figure 12 yields an estimate of total calendar-year wind turbine equipment imports – including wind-powered generating sets as shown in Figure 11 and selected turbine components as shown in Figure 12 – into the United States of $2.5 billion in 2006, $4.6 billion in 2007, $5.4 billion in 2008, and $4.2 billion in 2009. In aggregate, imports have substantially increased over time, peaking in 2008 and then declining in 2009. Annual wind power capacity additions have, however, outpaced imports over this timeframe, suggesting a growing share of domestic production.

To estimate the percentage share of imports and domestic production over time, one must account for the fact that turbines and components imported at the end of one year may not be installed until the following year. As such, in Figure 13 we determine the combined imports of wind-powered generating and selected turbine components by using a 4-month lag (i.e., we use monthly import data from September of the previous year to August of the current year to estimate the value of imports used in wind turbine installations in the current year). Those import figures are then compared to total wind turbine equipment-related costs on a calendar-year basis.[27] When presented as a fraction of total equipment-related turbine costs in this fashion, the overall import fraction is found to have declined significantly from more than 80% in 2006 to roughly 40% in 2009.[28]

These figures should be considered rough approximations for the reasons stated earlier, and may understate the wind power industry's reliance on turbine and component imports because it is possible that imports of wind power equipment are occurring under other HTS codes that are not captured here. Nonetheless, the overall findings presented here are directionally consistent with the data presented in AWEA (2010a) and David (2010): a growing amount of the equipment used in wind power projects is being sourced domestically as domestic and foreign companies seek to minimize transportation costs and currency risks through local manufacturing. Domestic manufacturing and foreign direct investment in the U.S. market are becoming increasingly prevalent relative to cross-border trade in wind power equipment (see also Kirkegaard et al. 2009). In fact, though imperfect, this analysis suggests a greater domestic share of turbine and component manufacturing than estimated by AWEA (2010a). Moreover, the planned manufacturing investments discussed in the previous section may lead to increased domestic content in the years ahead: whether that trend continues unabated, however, may depend on the size and stability of the U.S. wind power market as well as the manufacturing strategies of emerging turbine manufacturers from Asia and elsewhere.

The Average Nameplate Capacity, Hub Height, and Rotor Diameter of Installed Wind Turbines Increased

The average nameplate capacity of wind turbines installed in the United States in 2009 increased to roughly 1.74 MW (Figure 14), up from 1.66 MW in 2008 and 1.65 MW in 2007.[29] Since 1998-99, average turbine nameplate capacity has increased by 145%, but growth in this metric has slowed in recent years due to the dominance of GE's 1.5 MW turbine and as a result of the logistical challenges associated with transporting larger turbines to project sites.[30]

Table 6. Size Distribution of Number of Turbines Over Time

	Years:	**1998-99**	**2000-01**	**2002-03**	**2004-05**	**2006**	**2007**	**2008**	**2009**
	# MW:	1,016	1,758	2,125	2,803	2,454	5,249	8,350	9,994
	# turbines:	1,425	1,987	1,757	1,960	1,536	3,190	5,029	5,734
Turbine Size Range (MW)	**0.00-0.5**	1.3%	0.4%	0.5%	1.8%	0.7%	0.0%	0.5%	0.2%
	0.51-1.0	98.5%	73.9%	43.4%	18.5%	10.7%	11.2%	10.3%	4.6%
	1.01-1.5	0.0%	25.4%	43.5%	56.0%	54.0%	49.2%	53.5%	49.4%
	1.51-2.0	0.3%	0.4%	12.5%	23.6%	18.4%	23.1%	16.3%	21.2%
	2.01-2.5	0.0%	0.0%	0.0%	0.1%	16.2%	15.2%	16.8%	23.1%
	2.51-3.0	0.0%	0.0%	0.1%	0.0%	0.0%	1.3%	2.5%	1.4%

Source: AWEA project database

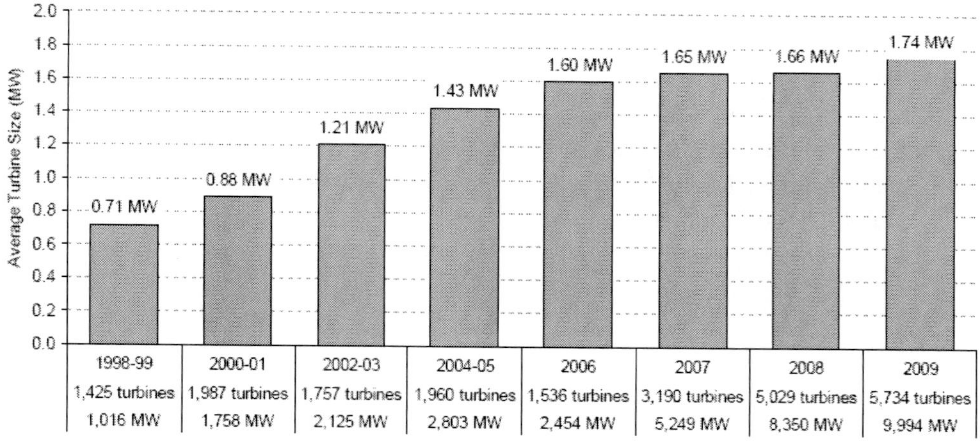

Source: AWEA project database

Figure 14. Average Turbine Nameplate Capacity Installed During Period.

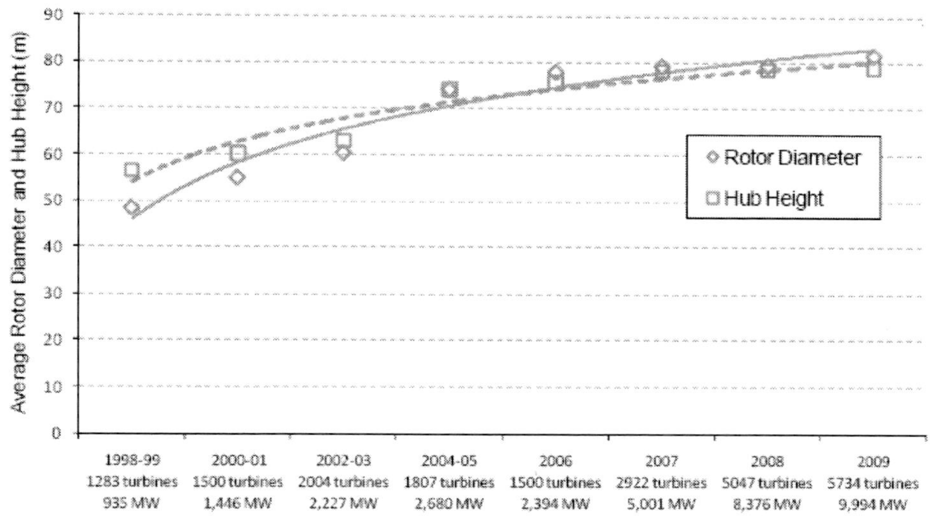

Source: Berkeley Lab

Figure 15. Average Rotor Diameter and Hub Height Installed During Period.

Table 6 shows how the distribution of turbine nameplate capacity has shifted over time: roughly 25% of all turbines installed in 2009 had a nameplate capacity larger than 2.0 MW, compared to 19% in 2008, 16% in both 2007 and 2006, and just 0.1% or less in years prior to 2006. GE's 1.5 MW wind turbine remained by far the nation's most-popular turbine in 2009, with 2,663 units installed, equating to 40% of all wind power capacity installed in 2009.

In addition to nameplate capacity ratings, average hub heights and rotor diameters have also scaled with time. The average hub height of wind turbines installed in the United States in 2009 was 78.8 meters (Figure 15), up slightly from 78.5 meters in 2008 and 78.2 meters in 2007. Since 1998-99, the average turbine hub height has increased by 39% (or 22.3 meters), though year-on-year growth has slowed in the more recent years. Average rotor diameters have increased at a somewhat more rapid pace: the average rotor diameter of wind turbines installed in the United States in 2009 was 81.6 meters (Figure 15), up from 79.4 meters in 2008 and 79.2 meters in 2007. Since 1998-99, the average rotor diameter has increased by 69% (or 33.2 meters). For turbines installed in 2009, the maximum hub height and rotor diameter were 80 meters and 101 meters, respectively (a higher maximum hub height of 105 meter exists for turbines installed in 2008). These trends in hub height and rotor scaling are one of several factors impacting the project-level capacity factors highlighted later in this report.

The Average Size of Wind Power Projects Resumed its Upward Trend

As the wind power industry has grown, so too has the average size of installed wind power projects. Projects installed in 2009 averaged nearly 91 MW, which is below the 120 MW average size of projects built in 2007, but is otherwise larger than in any other period (Figure 16).

The long-term increase in average project size may reflect a number of interrelated trends highlighted elsewhere in this report: growing demand for wind power; the upward march in turbine size; the large turbine orders that had become standard practice up until the 200 8/2009 credit crisis; consolidation among project developers to support those orders; and increasing turbine and project costs, which may require taking full advantage of any and all economies of scale. Whatever the specific cause, larger project sizes reflect an increasingly mature energy source that is beginning to penetrate into the domestic electricity market in a significant way.

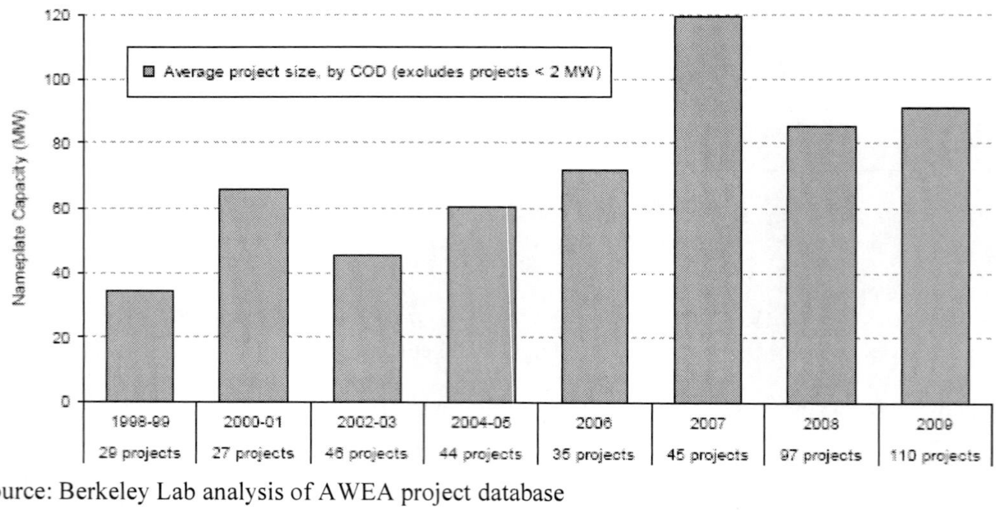

Source: Berkeley Lab analysis of AWEA project database

Figure 16. Average Project Size, by Commercial Operation Date (COD) [31].

Consolidation among Wind Project Developers Continues

Consolidation on the development end of the wind power business has slowed somewhat since 2007, but in 2009 remained on par with the pace set in 2008. The more-subdued pace of activity since 2007 may be a reflection of several factors, including the simple fact that many of the prime targets for investment and/or acquisition were acquired in earlier years. In addition, some traditional buyers of U.S. wind assets may have decided to reign in new investments following aggressive purchases made in previous years, while some developers who might otherwise entertain offers may be holding out for better pricing as the market recovers. Looking ahead, however, the relatively weak demand for wind energy projected in 2010 (and the trouble that will cause for smaller developers), coupled with an influx of cash from the Section 1603 Treasury grant program, may help to drive continued consolidation.

Table 7 provides a listing of announced acquisition and investment activity among U.S. wind project developers from 2002 through 2009.[32] At least six significant transactions involving roughly 18 GW of in-development wind power projects (also called the development "pipeline") were announced in 2009, similar to the five transactions and 19 GW in 2008, but well below the 11 transactions and 37 GW in 2007, and the 12 transactions and 34 GW in 2006. In 2005, eight transactions totaling 11 GW were announced, while only four transactions totaling less than 4 GW were completed from 2002 through 2004.

Table 7. Announced Acquisition and Investment Activity Among Wind Developers*

Investor	Transaction Type	Developer	Announcement Date
EDF (SIIF Energies)	Acquisition	enXco	May-02
Gamesa	Investment	Navitas	Oct-02
AES	Investment	U.S. Wind Force	Sep-04
PPM (Scottish Power)	Acquisition	Atlantic Renewable Energy Corp.	Dec-04
AES	Acquisition	SeaWest	Jan-05
Goldman Sachs	Acquisition	Zilkha (Horizon)	Mar-05
JP Morgan Partners	Investment	Noble Power	Mar-05
Arclight Capital	Investment	CPV Wind	Jul-05
Diamond Castle	Acquisition	Catamount	Oct-05
Pacific Hydro	Investment	Western Wind Energy	Oct-05
EIF U.S. Power Fund II	Investment	Tierra Energy, LLC	Dec-05
Airtricity	Acquisition	Renewable Generation Inc.	Dec-05
Babcock & Brown	Acquisition	G3 Energy LLC	Jan-06
Iberdrola	Acquisition	Community Energy Inc.	Apr-06
Shaw/Madison Dearborn	Investment	UPC Wind	May-06
NRG	Acquisition	Padoma	Jun-06
CPV Wind	Acquisition	Disgen	Jul-06

Table 7. (Continued)

Investor	Transaction Type	Developer	Announcement Date
BP	Investment	Clipper	Jul-06
BP	Acquisition	Greenlight	Aug-06
Babcock & Brown	Acquisition	Superior	Aug-06
Enel	Investment	TradeWind	Sep-06
Iberdrola	Acquisition	Midwest Renewable Energy Corp.	Oct-06
Iberdrola	Acquisition	PPM (Scottish Power)	Dec-06
BP	Acquisition	Orion Energy	Dec-06
Naturener	Acquisition	Great Plains Wind & Energy, LLC	Feb-07
HSH Nordbank	Investment	Ridgeline Energy	Feb-07
Energias de Portugal	Acquisition	Horizon	Mar-07
Iberdrola	Acquisition	CPV Wind	Apr-07
Duke Energy	Acquisition	Tierra Energy, LLC	May-07
Acciona	Acquisition	EcoEnergy, LLC	Jun-07
Babcock & Brown	Acquisition	Bluewater Wind	Sep-07
Good Energies	Investment	EverPower	Sep-07
E.ON AG	Acquisition	Airtricity North America	Oct-07
Wind Energy America	Acquisition	Boreal	Oct-07
Marubeni	Investment	Oak Creek Energy Systems	Dec-07
NTR	Investment	Wind Capital Group	Apr-08
Canadian Pension Plan	Investment	Noble Power	Apr-08
ArcLight and Terra-Gen	Acquisition	Allco Wind Energy	Jun-08
Duke Energy	Acquisition	Catamount	Jun-08
Veolia	Acquisition	Ridgeline Energy	Oct-08
Riverstone Holdings	Acquisition	Babcock & Brown	Jun-09
Terra Firma	Acquisition	Everpower Wind	Aug-09
APEX Wind Energy	Acquisition	BQ Energy, LLC	Jun-09
Global Infrastructure Partners	Investment	Terra-Gen Power Holdings	Nov-09
NRG Energy	Acquisition	Bluewater Wind	Nov-09
Enel	Investment	Geronimo Wind	Nov-09

* Select list of announced transactions; excludes joint development activity
Source: Berkeley Lab.

A number of large companies have entered the U.S. wind project development business in recent years, some through acquisitions and investments as highlighted in Table 7, and others through their own development activity or through joint development agreements with others. Particularly striking in recent years has been the entrance of large European energy companies, as well as the increased interest of U.S. utility affiliates in wind project development.

Treasury Cash Grant Expands Financing Options, Buoys the Wind Sector

Due to the global credit crisis, wind power project financing in the United States declined precipitously at the close of 2008, with many lenders and tax equity investors sidelined by extreme uncertainty and shrinking tax capacity. By mid-February of 2009, however, the U.S. Congress had passed the Recovery Act, parts of which were intended to alleviate financial constraints on the industry. Most notably, Section 1603 of the Recovery Act enables wind (and other qualifying) power projects to temporarily choose a 30% cash grant administered by the U.S. Treasury in lieu of either the PTC or a 30% investment tax credit (ITC). Title IV of the Recovery Act also expands an existing federal loan guarantee program administered by DOE to renewable energy projects using commercially proven (rather than just innovative) technology.

By replacing the PTC with an up-front 30% cash grant, Section 1603 greatly reduces, but does not completely eliminate, the dependence of wind project developers on third-party tax equity investors. Tax appetite from outside of the project itself is still needed to efficiently use accelerated depreciation deductions in the year that they are generated. Many developers, however, have found that financing their projects with low-cost debt rather than more-expensive tax equity can more-than-make-up-for the value lost from carrying forward depreciation deductions until they can be fully used within the project itself.

The Section 1603 cash grant program has been heavily subscribed by the industry. Owners of more than 6,400 MW – i.e., more than 64% – of the wind power capacity installed in 2009 elected the grant in lieu of the PTC. As much as 2,400 MW of this capacity may not have been built in 2009 had the cash grant not been available (Bolinger et al. 2010). And in another sign that Section 1603 has accomplished its goal of reducing dependence on the tax equity market, only about seven of the more-than-sixty 2009 projects that elected the grant were financed using third-party tax equity (Chadbourne & Parke 2010a); many of the rest substituted project-level term debt for third-party tax equity (and are presumably planning to carry forward unused depreciation deductions), while still others were built by developers that have their own internal tax appetite.

At present, wind power projects must be under construction by the end of 2010, and online by the end of 2012, in order to qualify for the grant. Congress is considering several bills that would extend the grant program in some form or fashion, and one oft-cited rationale for an extension is that the tax equity market has not yet recovered sufficiently to supply the amount of capital that the market would otherwise require. As of May 2010, there were roughly a dozen tax equity investors active in the wind power market – up from the handful of investors that maintained their presence throughout 2009, but still down from the market heights of early 2008 (Stolarski et al. 2010, Martin 2010). That said, as of June 2010, more than $2 billion of tax equity had reportedly been invested in the wind power sector since the start of the year –more than was invested throughout all of 2009 – and the market was considered to be on pace to reach tax equity investment levels seen in the peak year of 2007 (Chadbourne & Parke 2010b).

Although the tax equity market may still be somewhat constrained, capital has been flowing more-freely in the debt market, both from banks and institutional lenders. So-called "mini- perms" (i.e., term debt with a balloon payment due in 5-7 years) dominated for much of 2009, though tenors began to lengthen somewhat as the year progressed. By early 2010, fully amortizing loans were once again seen in the market, some with tenors as long as 15-17 years. Spreads remain high – as high as 350 basis points above LIBOR in early 2010, compared to just 125 basis points a few years ago – but were reportedly under downward pressure in mid-2010 as the number of banks active in the wind power market increased to more than thirty (Zaelke et al. 2010), and in fact had fallen below 300 basis points according to some sources (Chadbourne & Parke 2010b).

The rebound in the debt market is just one reason that there has been relatively weak demand for the federal loan guarantee program that was created as part of the *Energy Policy Act of 2005* and expanded by the Recovery Act to include projects using commercially proven technologies. Other oft-cited reasons that the Section 1703 (for projects using innovative technology) and Section 1705 (for projects using commercially proven technology) loan guarantee programs have not been more popular include the relatively slow initial implementation of these programs, challenging application requirements (including the need to for a federal environmental review and complying with the *Davis-Bacon Wage Act*), initial inflexibility with regard to financing structures involving third-party tax equity, and the additional complexities and time to close that come from having another party – DOE – at the bargaining table (Bailey 2010, Zaelke et al. 2010, Stolarski et al. 2010, Chadbourne & Parke 2010b). By mid-July 2010, just two wind- related loan guarantees had been awarded under the Section 1703 program: Nordic Windpower received a $16 million loan guarantee to expand its wind turbine manufacturing facility in Pocatello, Idaho, while First Wind received a $117 million loan guarantee for its 30 MW Kahuku wind project (which includes battery storage) in Hawaii. No wind power project has yet been awarded a guarantee under the Section 1705 program for commercial technologies, but that program could be particularly useful for larger wind power projects that can spread the transaction costs over more capacity, and that might otherwise be too large to raise debt financing through the normal channels (Zaelke et al. 2010, Stolarski et al. 2010, Chadbourne & Parke 2010b).

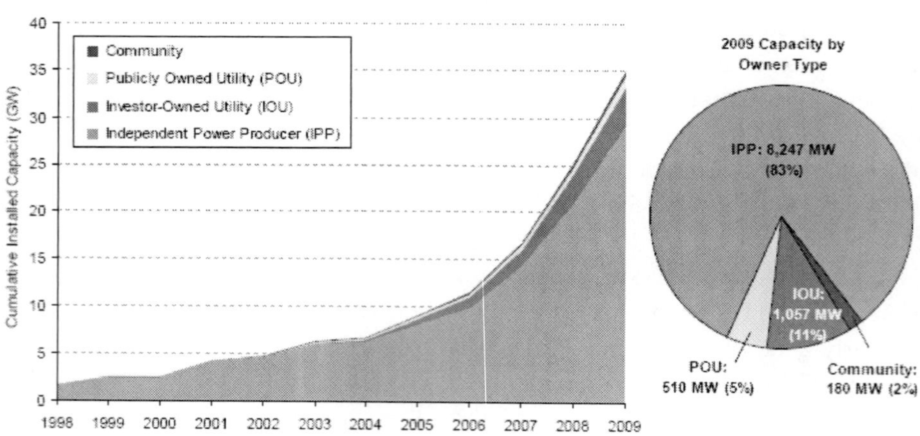

Source: Berkeley Lab estimates based on AWEA project database

Figure 17. Cumulative and 2009 Wind Power Capacity Categorized by Owner Type.

IPP Project Ownership Remained Dominant, but Utility Ownership Increased

Independent power producers (IPPs) continued to dominate the ownership of wind power projects in 2009, owning 83% (8,247 MW) of all new capacity additions (Figure 17). Nearly 16% of the total wind power capacity additions in 2009 are owned by local electric utilities, with investor-owned utilities (IOUs) owning 1,057 MW and publicly owned utilities (POUs) owning another 510 MW. Community wind power projects – defined here as projects using turbines over 100 kW in size and completely or partly owned by towns, schools, commercial customers, or farmers, but excluding publicly owned utilities – constitute the remaining 2% of new capacity, with 180 MW. Of the cumulative installed wind power capacity at the end of 2009, IPPs owned 83% (29,164 MW), with utilities contributing 15% (4,265 MW for IOUs and 1,071 MW for POUs), and community ownership just 2% (656 MW).

The dominance of IPP ownership, and the more recent trend towards increased utility ownership, has been driven by several factors. Up until the Internal Revenue Service (IRS) clarified the issue in 2005, some IOUs were uncertain as to whether they could claim the PTC on utility- owned wind power projects (due to the requirement that PTC-eligible power must be sold to an unrelated party – in 2005 the IRS clarified that ratepayers are indeed unrelated parties). More broadly, when wind energy was a small part of the generation mix, some utilities felt that buying wind power was less risky than owning wind power projects. As utilities have gained comfort with wind power over the years, however, their interest in ownership has increased for several reasons: IOUs are typically allowed to earn a regulated return on project ownership (i.e., by adding it to their rate base) but not on power purchases; credit rating agencies have at times considered long-term power purchase agreements to be debt-like instruments, thereby potentially negatively impacting a utility's credit rating; and ownership places the utility in a position of greater control over the cost and price of wind energy that it receives. As a result of these drivers, utility ownership of wind power projects may continue to increase in the coming years.

Long-Term Contracted Sales to Utilities Remained the Most Common Sales Arrangement, but Merchant Plants Were Surprisingly Abundant in 2009

Electric utilities continued to be the dominant purchasers of wind power (see Figure 18), with 58% of the new 2009 capacity selling electricity under long-term power purchase agreements (PPAs) to either IOUs (36%) or POUs (22%). In aggregate, these two types of utilities buy power from 62% of the cumulative wind power capacity in the United States (IOUs purchase 44% and POUs purchase 18%).

Surprisingly, merchant/quasi-merchant projects were also in abundance in 2009, accounting for 38% of all new capacity (and 26% of cumulative capacity). Merchant/quasi-merchant projects are those whose electricity sales revenue is tied to short-term contracted and/or wholesale spot electricity market prices (with the resulting price risk commonly hedged over a 5- to 10-year period[33]) rather than being locked in through a long-term PPA. Expectations for merchant development in 2009 had been low at the start of the year, due to very tight credit and sharply lower wholesale electricity prices. In fact, it is possible that many projects that sold power on a merchant basis in 2009 may now be seeking longer-term PPAs in order to gain increased revenue stability. Regardless, the amount of merchant/quasi-merchant sales in 2009 was significant, with approximately 72% of that activity in Texas (45%), New York (15%), and Illinois (12%) – i.e., states in which wholesale spot electricity markets exist, where wind power has (at least prior to 2009) been able to compete with those prices, and where additional revenue may be possible from the sale of renewable energy certificates (RECs).

The role of power marketers – defined here as corporate intermediaries that purchase power under contract and then re-sell that power to others, sometimes taking some merchant risk[34] – in the wind power market has waned somewhat in recent years. In 2009, power marketers purchased the output of just 4% of the new wind power capacity, with 11% of the cumulative capacity selling to power marketers.

Finally, roughly 50 MW of the wind power additions in 2009 that used turbines over 100 kW in size were interconnected on the customer side of the utility meter, with the power being consumed on site rather than sold.

4. Price, Cost, and Performance Trends

Upward Pressure on Wind Power Prices Continued in 2009

Although some of the cost pressures facing the industry in recent years (e.g., rising materials costs, the weak dollar, turbine and component shortages) have eased somewhat, it will take time before relief flows through the project development pipeline to impact overall average wind power prices. After all, projects built in 2009 may have purchased turbines in 2007 or 2008, and may have established contractual pricing terms at a similar point in time. As such, 2009 was another year of rising wind power prices.

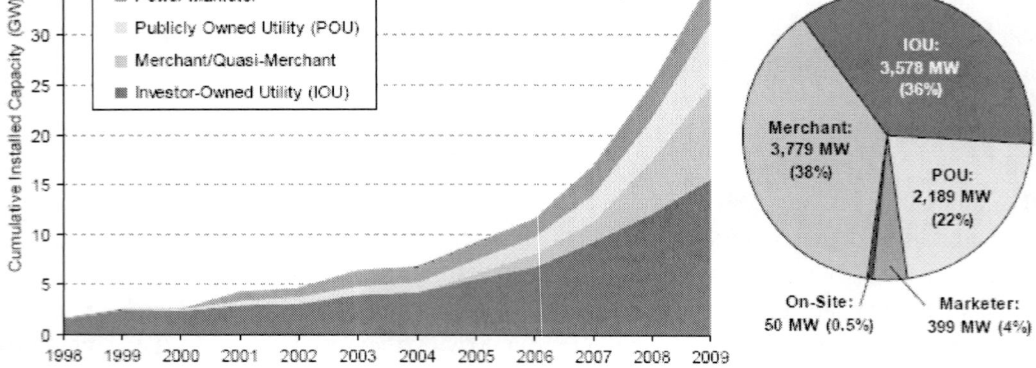

Source: Berkeley Lab estimates based on AWEA project database

Figure 18. Cumulative and 2009 Wind Power Capacity Categorized by Power Off-Take Arrangement.

Berkeley Lab collects data on wind power sales prices from the sources listed in the Appendix, resulting in a dataset that consists of price data for 180 wind power projects installed between 1998 and the end of 2009. These projects total 12,813 MW, or 38% of the wind power capacity brought on line in the United States over the 1998-2009 timeframe.[35] The dataset excludes merchant plants and projects that sell renewable energy certificates (RECs) separately. The prices in the dataset therefore reflect the bundled price of electricity and RECs as sold by the project owner under a power purchase agreement. Because these prices are suppressed by the receipt of available state and federal incentives (e.g., the prices reported here would be at least $20/MWh higher without the PTC / ITC / Treasury Grant), they do not represent wind energy generation *costs*.

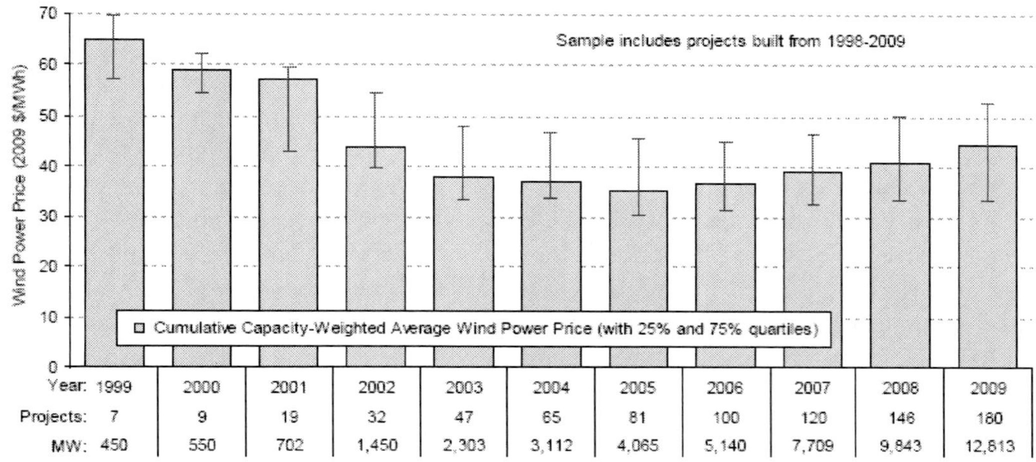

Source: Berkeley Lab

Figure 19. Cumulative Capacity-Weighted Average Wind Power Prices Over Time.

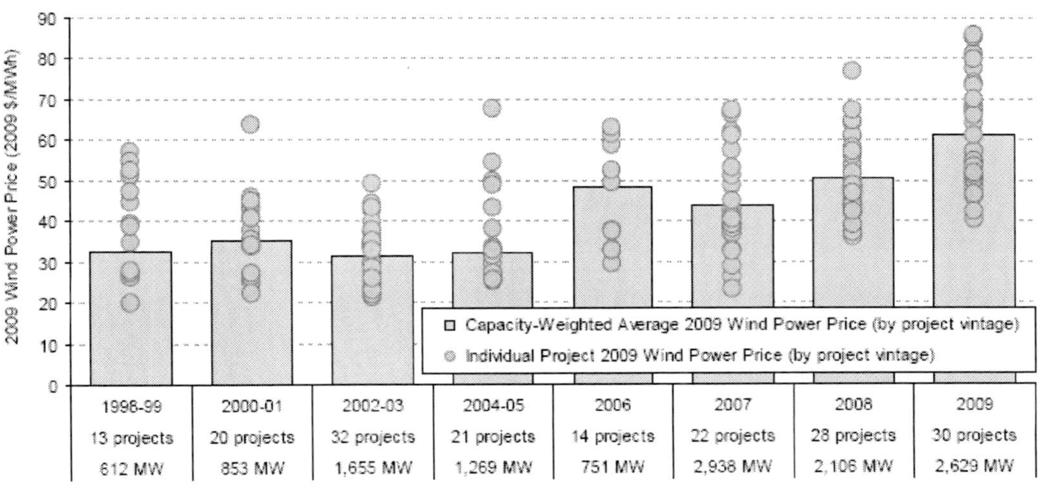

Source: Berkeley Lab database

Figure 20. 2009 Wind Power Prices by Project Vintage.

Based on these data, the capacity-weighted average power sales price from the sample of post- 1997 wind power projects remains relatively low by historical standards, but has been steadily increasing in recent years. Figure 19 shows the cumulative capacity-weighted average wind power price (along with the range of individual project prices falling between the 25th and 75th percentiles) in each calendar year from 1999 through 2009. Based on the limited sample of 7 projects built in 1998 or 1999 and totaling 450 MW, the weighted-average price of wind energy in 1999 was $65/MWh (expressed in 2009 dollars). By 2009, in contrast, the cumulative sample of projects built from 1998 through 2009 had grown to 180 projects totaling 12,813 MW, with an average price of $45/MWh (with 50% of individual project prices falling between $33/MWh and $53/MWh).[36] Although Figure 19 does show a modest increase in the weighted-average wind power price since 2005, reflecting rising prices from new projects, the cumulative nature of the graphic mutes the degree of increase.

To better illustrate changes in the price of power from newly built wind power projects, Figure 20 shows average wind power sales prices in 2009, grouped by project vintage (i.e., by each project's initial commercial operation date).[37] Although the limited project sample and the considerable variability in prices across projects installed in a given time period complicate analysis of national price trends (with averages subject to regional and other factors), the general trend exhibited by the capacity-weighted-average prices (i.e., the blue columns) nevertheless shows that prices bottomed out for projects built in 2002 and 2003, and have since risen significantly.[38] Specifically, the capacity-weighted average 2009 sales price, based on projects in the sample built in 2009, was roughly $61/MWh, up from an average of $51/MWh for the sample of projects built in 2008, and nearly double the average of $32/MWh among projects built during the low point in 2002 and 2003.

The underlying variability in wind power prices within a year is caused in part by regional factors, which may affect not only project capacity factors (depending on the strength of the wind resource in a given region), but also development and installation costs (depending on a region's physical geography, population density, labor rates, or even regulatory processes). It is also possible that regions with higher wholesale electricity prices or with greater demand for renewable energy will, in general, yield higher wind energy contract prices due to market factors.

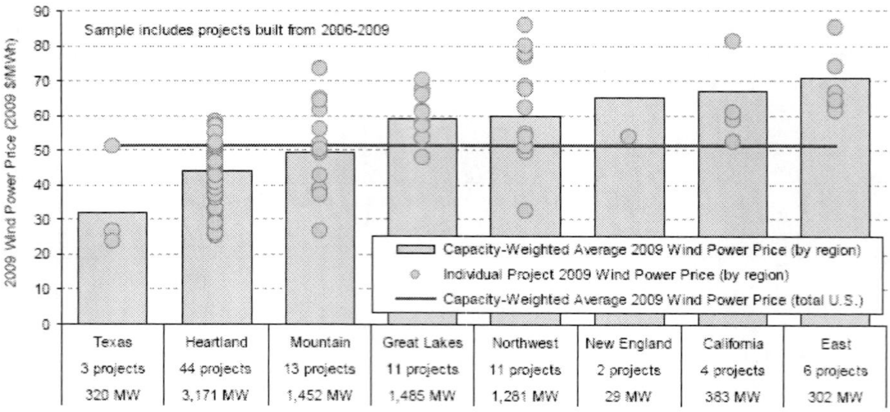

Source: Berkeley Lab

Figure 21. 2009 Wind Power Prices by Region: 2006-2009 Projects Only.

Figure 21 shows individual project and average 2009 wind power prices by region for just those wind power projects installed from 2006-2009 (i.e., the more-recent period of higher prices, as shown in Figure 20), with regions as defined in Figure 22. Although sample size is quite small and therefore problematic in numerous regions, Texas and the Heartland region appear to be among the lowest price areas on average, while New England, California, and the East are among the higher price regions.[39]

Sharp Drop in Wholesale Electricity Prices Makes the Near-Term Economics of Wind Energy More Challenging

A simple comparison of the wind power prices presented in the previous section to recent wholesale electricity prices throughout the United States demonstrates that while wind power had consistently been priced (on average) at the low end of the range of wholesale electricity prices going back through 2003, the drop in wholesale electricity prices in 2009 pushed wind energy to the top of that range. Specifically, Figure 23 shows the range (minimum and maximum) of average annual wholesale electricity prices for a flat block of power[40] going back to 2003 at twenty-three different pricing nodes located throughout the country (refer to Figure 22 for the names and approximate locations of the twenty-three pricing nodes represented by the blue-shaded area[41]). The red dots show the cumulative capacity-weighted average price received by wind power projects in each year among those projects in the sample with commercial operation dates of 1998 through 2009 (consistent with the data first presented in Figure 19).

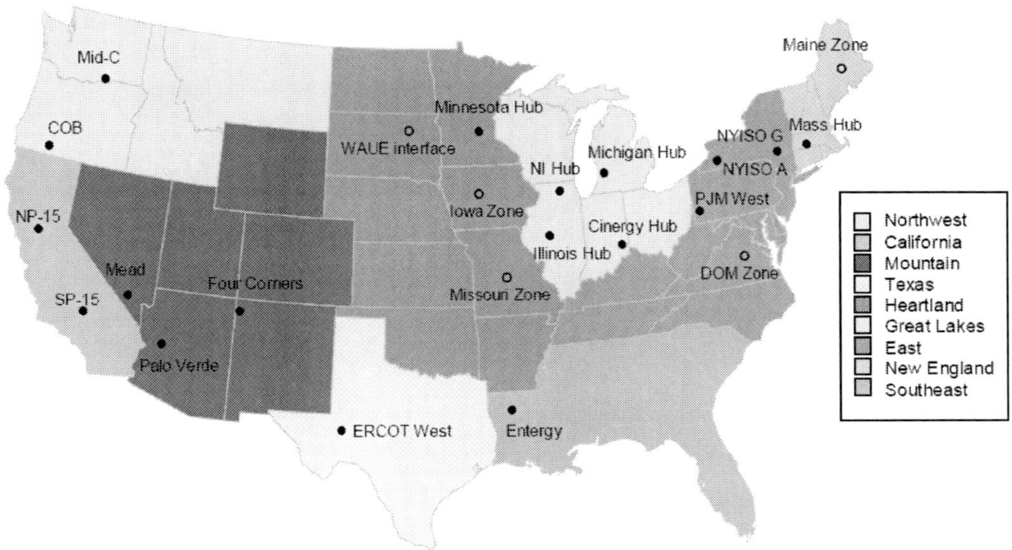

Note: The pricing nodes represented by an open, rather than closed, bullet do not have complete pricing history back through 2003.

Figure 22. Map of Regions and Wholesale Electricity Price Hubs Used in Analysis.

REC Markets Remain Fragmented, with a Wide Range of Pricing

The wind power sales prices presented in this report reflect only the bundled sale of both electricity and RECs; excluded are projects that sell RECs separately from electricity, thereby generating two sources of revenue. REC markets are highly fragmented in the United States, but consist of two distinct segments: compliance markets in which RECs are purchased to meet state RPS obligations, and green power markets in which RECs are purchased on a voluntary basis.

The figures below present indicative monthly data of spot-market REC prices in both compliance and voluntary markets, grouped into High-Price and Low-Price markets; data for compliance markets focus on the "Class I" or "Main Tier" of the RPS policies. Clearly, spot REC prices have varied substantially, both among states and over time within individual states. Among compliance markets in the Northeast, prices for RECs used to serve RPS requirements in Connecticut, Massachusetts, and Rhode Island remained relatively flat in 2009, following a steep drop in 2008, while prices in the new RPS compliance markets in New Hampshire and Maine were at levels consistent with the other Northeastern states. REC prices to serve RPS requirements in New Jersey, Illinois, and Delaware remained relatively flat during the latter half of the 2009, after declining earlier in the year. REC prices remained relatively low in several other compliance markets (Texas, Maryland, Pennsylvania, and Washington D.C.) due to a surplus of eligible renewable energy supply relative to RPS-driven demand in those markets. Prices for RECs offered in voluntary markets in 2009 ranged from an annual average of less than \$2/MWh for national voluntary wind RECs (which continue to closely track the price of Texas RECs) to approximately \$7/MWh for voluntary wind RECs in the West.

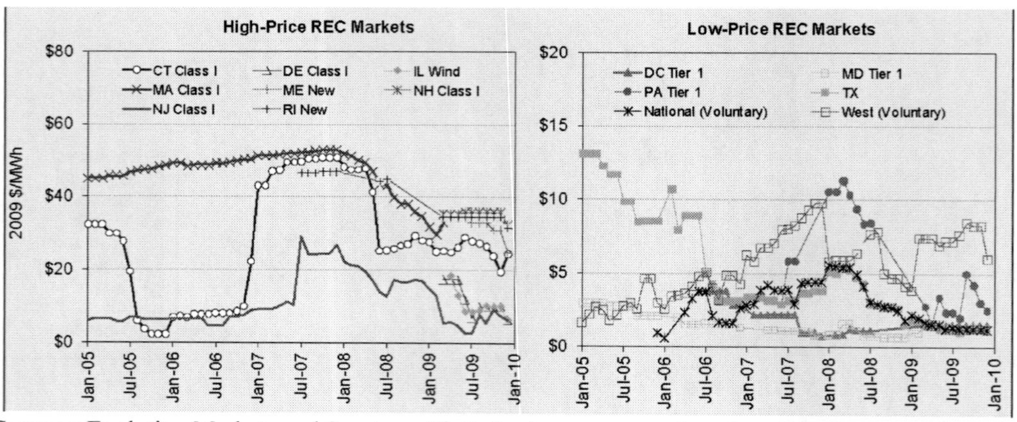

Sources: Evolution Markets and Spectron. Plotted prices represent the price of the last monthly trade (if available), or the midpoint of Bid and Offer prices, for the current or nearest compliance year.

At least on a cumulative basis within the sample of projects reported here, average wind power prices compared favorably to wholesale electricity prices from 2003 through 2008. The increase in wind power prices in 2009, however, combined with the deep reduction in wholesale electricity prices (driven by lower natural gas prices), reversed this long-term trend in 2009. Although low natural gas prices are, in part, attributable to the recession-induced drop in energy demand, the discovery and early development of significant shale gas deposits has resulted in reduced expectations for increases in natural gas prices going forward. As a result, natural gas prices may not rebound to earlier levels as the economy recovers, putting the near-term comparative economic position of wind energy at some risk (especially if wind energy costs do not also decrease – see a later section suggesting that cost reductions may be on the horizon).

Though Figure 23 portrays a national comparison, there are clearly regional differences in wholesale electricity prices and in the average price of wind power. Moreover, as shown earlier in Figure 20, wind power prices have risen among more-recently built projects. Figure 24 accounts for both of these considerations by focusing on 2009 wind and wholesale electricity prices in the same regions as shown earlier, based only on the sample of wind power projects installed from 2006 through 2009 (i.e., the more-recent period of higher pricing, as shown earlier in Figure 20).[42] Although there is quite a bit of variability within some regions, and several regions again have limited sample size, the spread between the average wind power and wholesale electricity prices (i.e., the wind power premium) in each region is fairly consistent across the United States, suggesting that the struggle for wind energy to compete in 2009 on short-term economics alone was indeed a nationwide phenomenon.

Important Note: Notwithstanding the comparisons made in Figures 23 and 24, it should be recognized that neither the wind nor wholesale electricity prices presented in this section reflect the full social costs of power generation and delivery. Specifically, the wind power prices are suppressed by virtue of federal and, in some cases, state tax and financial incentives. Furthermore, these prices do not fully reflect integration, resource adequacy, or transmission costs. At the same time, wholesale electricity prices do not fully reflect transmission costs, may not fully reflect capital and fixed operating costs, and are suppressed by virtue of any financial incentives provided to fossil-fueled generation and by not fully accounting for the environmental and social costs of that generation. In addition, wind power prices – once established – are typically fixed and known (because wind energy is often sold through long-term, fixed-price power purchase agreements), whereas wholesale electricity prices are short-term and therefore subject to change over time. Finally, the location of the wholesale electricity nodes and the assumption of a flat-block of power are not perfectly consistent with the location and output profile of the sample of wind power projects.

In short, comparing wind and wholesale electricity prices in this manner is not appropriate if one's goal is to fully account for the costs and benefits of wind energy relative to its competition. Another way to think of Figures 23 and 24, however, is as loosely representing the decision facing wholesale electricity purchasers that are otherwise under no obligation to purchase additional amounts of wind energy – i.e., whether to contract long-term for wind power or to buy a flat block of (non-firm) spot power on the wholesale electricity market. In this sense, the costs represented in Figures 23 and 24 are reasonably comparable, in that they represent (to some degree, at least) what the power purchaser would actually pay.

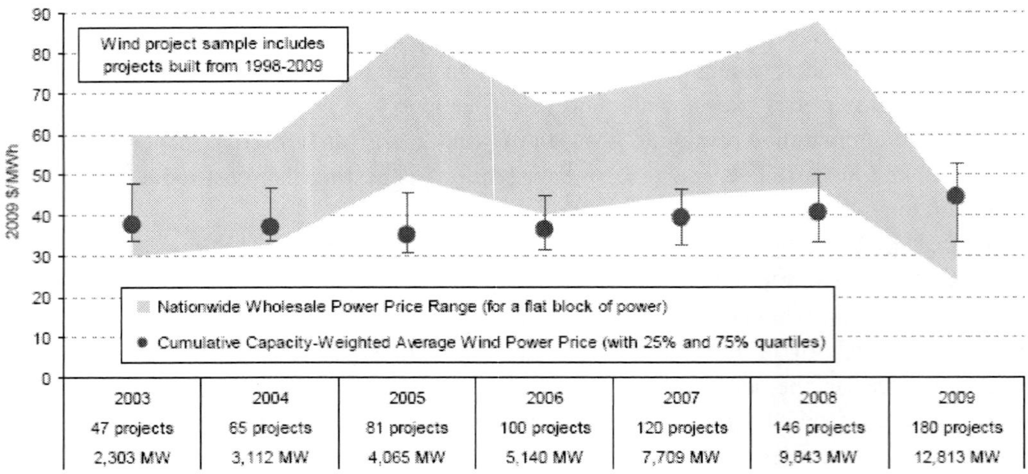

Source: Berkeley Lab, FERC, Ventyx, ICE

Figure 23. Average Cumulative Wind and Wholesale Electricity Prices Over Time.

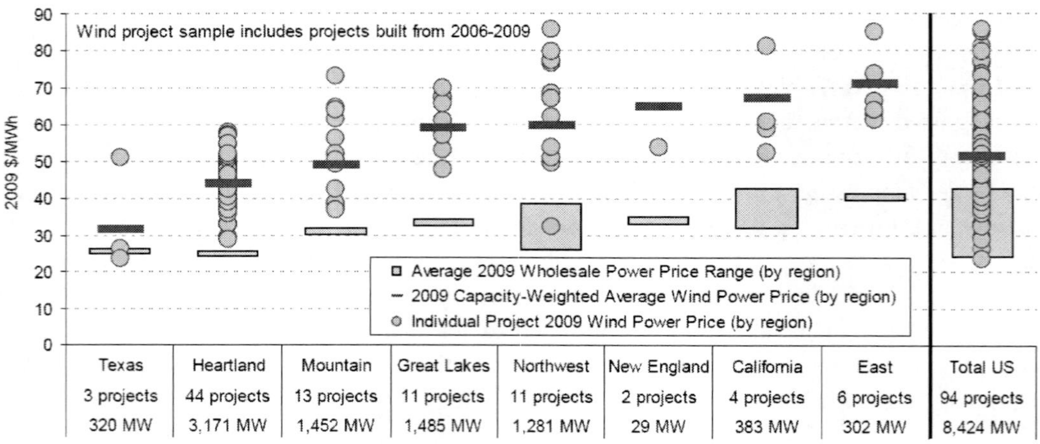

Source: Berkeley Lab, Ventyx, ICE

Figure 24. Wind and Wholesale Electricity Prices by Region: 2006-2009 Projects Only.

Project Performance and Capital Costs Drive Wind Power Prices

Wind power sales prices are affected by a number of factors, two of the most important of which are installed project costs and project performance.[43] Figures 25 and 26 illustrate the importance of these two variables.

Figure 25 shows the relationship between project-level installed costs and power sales prices in 2009 for a sample of more than 10,500 MW of wind power projects installed in the United States from 1998 through 2009.[44] Though the scatter is considerable, in general, projects with higher installed costs also have higher wind power prices.

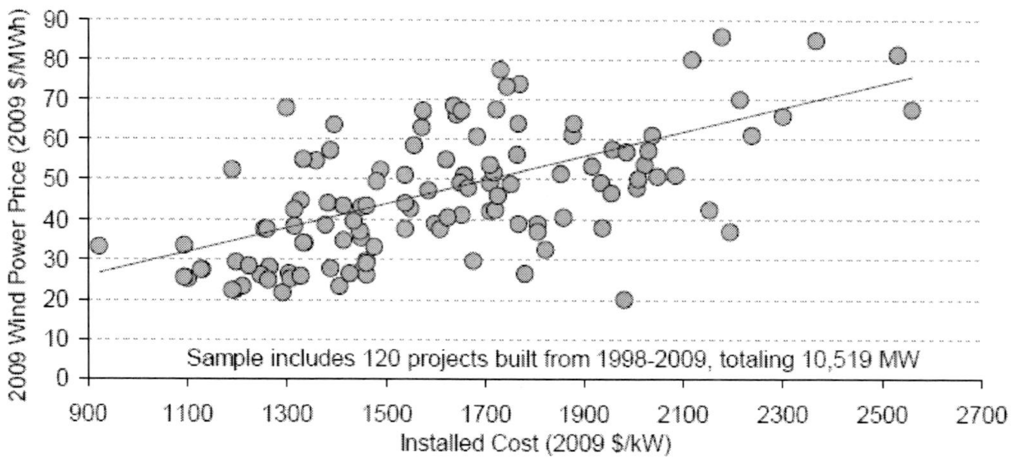

Source: Berkeley Lab

Figure 25. 2009 Wind Power Price as a Function of Installed Project Costs.

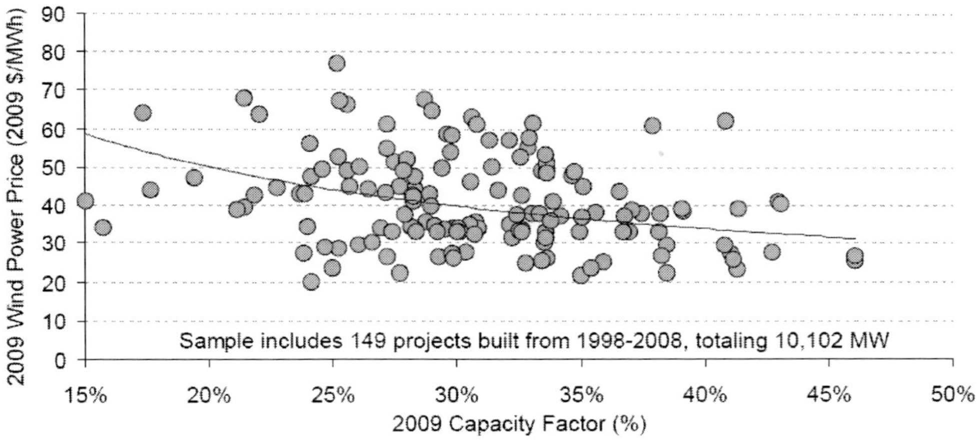

Source: Berkeley Lab

Figure 26. 2009 Wind Power Price as a Function of 2009 Capacity Factor.

Figure 26 illustrates the relationship between project-level capacity factors in 2009 and power sales prices in that same year for a sample of more than 10,100 MW of wind power projects installed from 1998 through 2008. The inverse relationship shows that projects with higher capacity factors generally have lower wind power prices, though considerable scatter is again apparent.

The next few sections of this report explore trends in installed costs and project performance in more detail, as both factors can have significant effects on wind power prices.

The Installed Cost of Wind Power Projects Continued to Rise in 2009, but Reductions May Be on the Horizon

Berkeley Lab compiles data on the installed cost of wind power projects in the United States, including data on 115 projects completed in 2009 totaling 9,656 MW, or 97% of the wind power capacity installed in that year. In aggregate, the dataset includes 405 completed wind power projects in the continental United States totaling 28,522 MW, and equaling roughly 81% of all wind power capacity installed in the United States at the end of 2009. In general, reported project costs reflect turbine purchase and installation, balance of plant, and any substation and/or interconnection expenses. Data sources are diverse, however, and are not all of equal credibility, so emphasis should be placed on overall trends in the data, rather than on individual project-level estimates.

As shown in Figure 27, the installed cost of wind power projects declined dramatically from the beginning of the industry in California in the 1980s through the early 2000s (falling by roughly $2,700/kW over this period[45]), but have more recently increased.[46] Among the sample of projects built in 2009, for example, the capacity-weighted average installed cost was $2,120/kW. This average increased by $170/kW (9%) from the weighted-average cost of $1,950/kW for projects installed in 2008, and increased by $820/kW (63%) from the average cost of projects installed from 2001 through 2004. Project costs have clearly risen, on average, over the last five years.[47]

Some of the cost pressures facing the industry in recent years (e.g., rising materials costs, the weak dollar, and turbine and component shortages) have eased since late 2008. As a result, while costs may – on average – remain high for a period of time as developers continue to work their way through the dwindling backlog of turbines purchased in early 2008 at peak prices under long-term frame agreements,[48] there are expectations that average installed costs will decline over time (see next section, on wind turbine price trends).

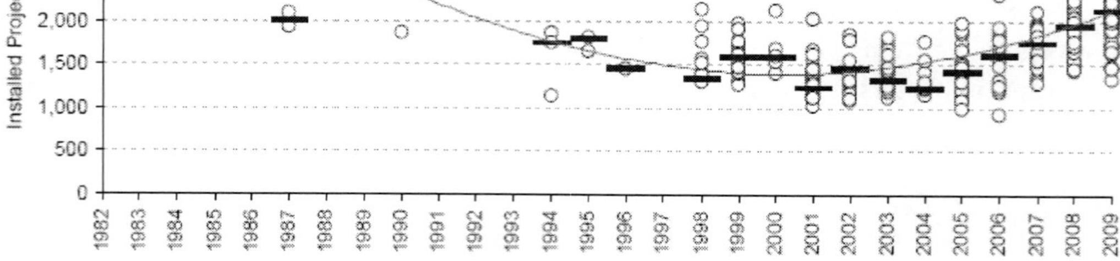

Source: Berkeley Lab (some data points suppressed to protect confidentiality)

Figure 27. Installed Wind Power Project Costs Over Time.

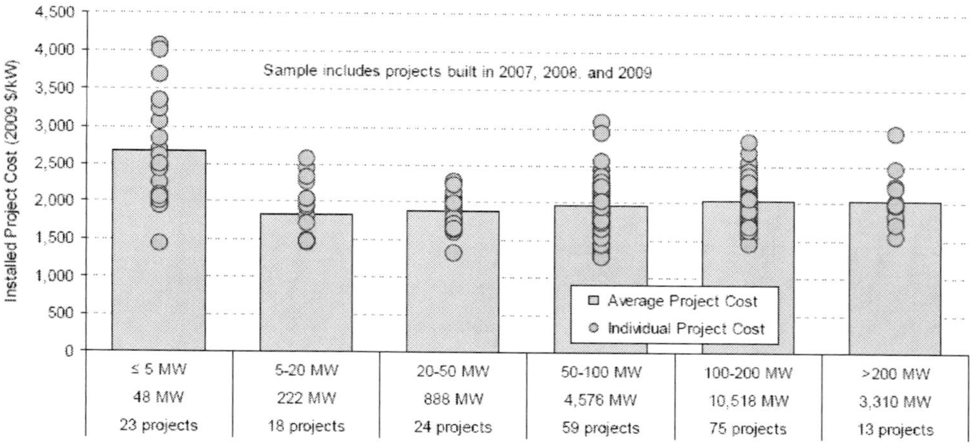

Source: Berkeley Lab

Figure 28. Installed Wind Power Project Costs by Project Size: 2007-2009 Projects.

Installed project costs exhibit economies of scale, at least at the low end of the project size range. Figure 28 shows that – among the sample of projects installed in 2007, 2008, or 2009 – there is a significant drop in per-kW average installed project costs when moving from projects of 5 MW or less to projects in the 5 to 20 MW range. As project size increases beyond 20 MW, these data do not show continued economies of scale; the reason for this latter trend is unclear.

Regional differences in average project costs are also apparent, and may occur due to variations in development costs, transportation costs, siting and permitting requirements and timeframes, and other balance-of-plant and construction expenditures. Considering only projects in the sample that were installed in 2007, 2008, and 2009, Figure 29 shows that the capacity-weighted average cost equaled $2,000/kW nationwide over this period. Texas was the lowest-cost region, while California and New England were the highest-cost regions; all other regions came in close to the nationwide average.[49]

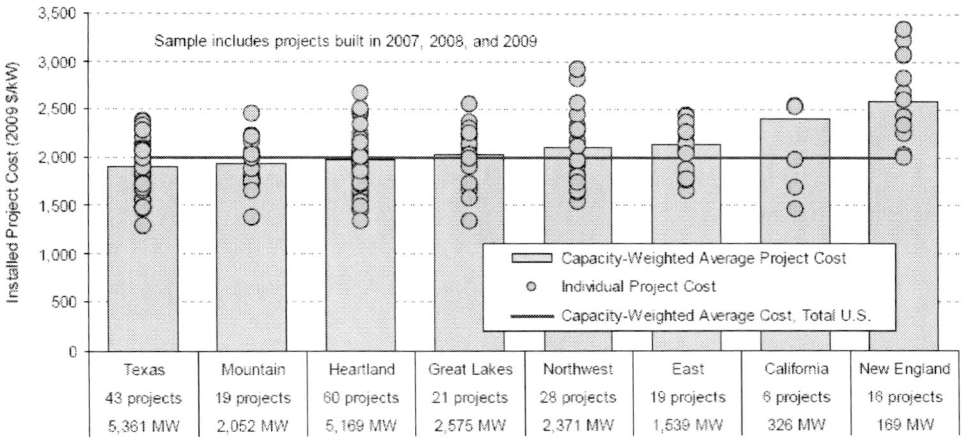

Source: Berkeley Lab

Figure 29. Installed Wind Power Project Costs by Region: 2007-2009 Projects.

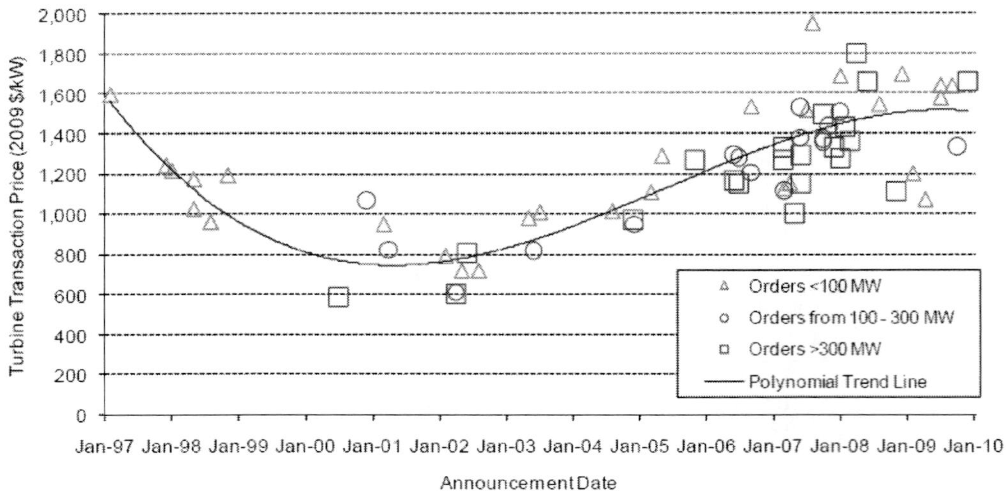

Source: Berkeley Lab

Figure 30. Reported U.S. Wind Turbine Transaction Prices Over Time.

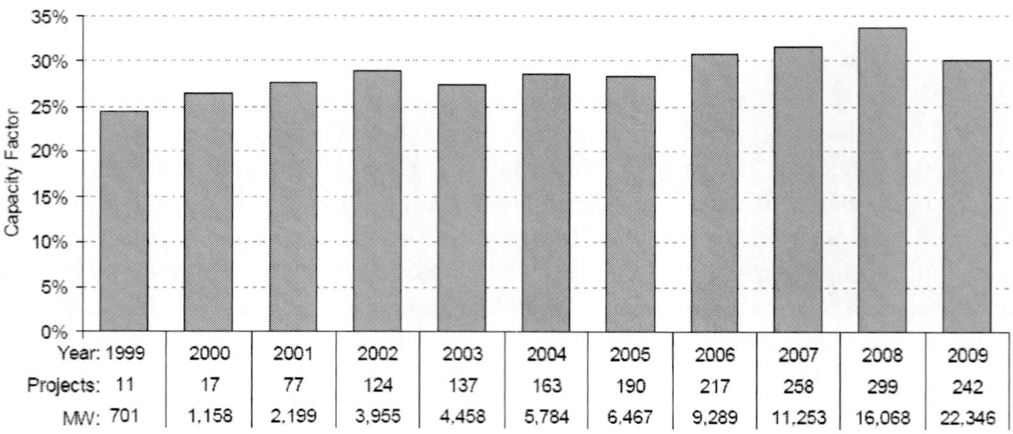

Source: Berkeley Lab

Figure 31. Average Cumulative Sample-Wide Capacity Factor by Calendar Year.

Wind Turbine Prices Have Begun to Show Signs of Easing, but Remain High By Historical Standards

Increases in wind power prices and overall installed project costs mirror increases in the cost of wind turbines over the last several years. Berkeley Lab has gathered data on 69 U.S. wind turbine transactions totaling 22,920 MW, including eight transactions summing to 1,674 MW announced in 2009. Figure 30 depicts these reported wind turbine transaction prices.

Sources of transaction price data vary, but most derive from press releases and news reports. Wind turbine transactions differ in the services offered (e.g., whether towers and installation are provided, the length of the service agreement, etc.) and on the timing of future turbine delivery, driving some of the observed intra-year variability in transaction prices. Nonetheless, most of the transactions included in the Berkeley Lab dataset likely include turbines, towers, erection, and limited warranty and service agreements.[50]

Since hitting a low point of roughly $700/kW in the 2000-2002 timeframe, average wind turbine prices have increased by approximately $800/kW (>100%) through 2009. The trend of increasing turbine prices also suggests that most of the rise in installed project costs reported earlier ($8 10/kW from 2001-04 through 2009) has come from turbine price increases. Increases in turbine prices over this period have been caused by several factors, including a decline in the value of the U.S. dollar relative to the Euro, increased materials and energy input prices (e.g., steel and oil), a general move by manufacturers to improve their profitability, shortages in certain turbine components, an up-scaling of turbine size (and hub height), and improved sophistication of turbine design (e.g., improved grid interactions).

Figure 30 also suggests that larger turbine orders may have generally yielded somewhat lower pricing than smaller orders at any given point in time. This is reflected in the fact that the majority of the largest turbine orders shown in Figure 30 are located below the polynomial trend line, while the majority of the smallest orders are located above that line.

Though turbine price increases have been the rule for a number of years, evidence is beginning to emerge that those days have ended, at least temporarily. As reflected by the small number of recent data points on turbine transactions shown in Figure 30, visibility of wind turbine transaction prices has declined as the financial crisis has taken its toll and developers sit on turbine supply frame agreements that have exceeded near-term project development plans. Energy and commodity prices have dropped since mid-2008, however, and the supply-demand balance for turbines has resulted in a turn towards a buyer's market (Bloomberg New Energy Finance 2010a; BTM 2010). As a result, UBS (2010) estimates a 13% decline in average turbine sales prices in 2009, while Bloomberg New Energy Finance (2010a) estimates that turbines delivered in the second half of 2010 are priced at a 15% discount relative to turbines delivered in the second half of 2008. More favorable terms are also on offer for turbine purchasers, including lengthier servicing agreements. These price reductions and improved terms can be expected, over time, to exert downward pressure on total project costs and wind power prices.

Wind Power Project Performance Has Generally Improved Over Time, but Has Leveled Off in Recent Years

Though turbine and installed project cost increases have driven wind power prices higher over the past several years, improvements in wind power project performance have mitigated these impacts to some degree. In particular, capacity factors have generally increased for projects installed more recently, driven by some combination of higher hub heights, larger rotor diameters, and other technological advancements. At the same time, these performance improvements appear to have leveled off in the most recent time period.

This section presents excerpts from a Berkeley Lab compilation of project-level capacity-factor data. The full data sample consists of 260 wind power projects built between 1983 and 2008, and totaling 22,366 MW (89% of nationwide installed wind power capacity at the end of 2008).[51]

Focusing on a progressively larger cumulative sample of projects in each calendar year,[52] Figure 31 demonstrates that average sample-wide wind power project capacity factors have, in general, gradually increased over time, from just over 24% in 1999 (for projects installed through 1998) to a high of nearly 34% in 2008 (for projects installed through 2007), before dropping to 30% in 2009 (for projects installed through 2008).

The general trend of increasing capacity factors shown in Figure 31 may be due to a combination of factors, including – most prominently – the increasing hub heights and rotor diameters of more recently completed projects (documented in an earlier section) that also increase fleet-wide average turbine size over time. Turbines with higher hub heights and with larger rotor diameters (relative to nameplate capacity) will tend to have higher average capacity factors.

The year-to-year variation in average capacity factors shown in Figure 31 – and especially the large drop in average capacity factors in 2009 – is also caused by changes in the quality of the wind resource from year to year and by wind power curtailment.

- **Wind Resource Variation:** In part as a result of El Niño, the year 2009, for example, was considered to be a generally poor wind year throughout much of the United States, with average wind speeds below their long-term average over much of the country (e.g., 3Tier 2010). The year 2008, meanwhile, was generally considered to be a good wind year. As a result, the large drop in average capacity factors between 2008 and 2009 is, in part, a reflection of natural yearly variations in national average wind resource conditions.
- **Wind Power Curtailment:** Increasing amounts of wind power curtailment in recent years also significantly reduced sample-wide average capacity factors in 2009. Curtailment of project output due primarily to transmission inadequacy (and, as a consequence, low wholesale electricity prices) is a growing problem, primarily in Texas, but also in other markets. Due to transmission inadequacy, wind power projects in West Texas (which represent a growing fraction of U.S. installations), for example, have been forced by grid operators to reduce their output (or have voluntarily chosen to do so in response to negative price signals in the wholesale electricity market). Figure 32 shows that roughly 17% of potential wind energy generation within ERCOT was curtailed in 2009, compared to 8% in 2008 and just 1% in 2007 (ERCOT 2010). Curtailment was also experienced, to a much lesser degree, in other regions. In MISO, for example, roughly 1% of potential wind energy output in 2009 was curtailed (MISO 2010). As shown in Figure 31, the national sample-wide average capacity factor in 2009 with curtailment was 30%. This sample-wide average capacity factor would have reached 32% if not for the curtailment experienced in ERCOT and MISO (and slightly higher were one to account for curtailment in other regions[53]).

Figure 33 (as well as Figure 34 and Table 8) presents capacity factor data in a different way, by focusing just on capacity factors in the year 2009, rather than in each calendar year.[54] Specifically, Figure 33 shows individual project as well as capacity-weighted average 2009 capacity factors broken out by each project's vintage (i.e., commercial operation date). The capacity-weighted average 2009 capacity factors in the Berkeley Lab sample increase from 21% for wind power projects installed before 1998 to roughly 26%-27% for projects installed from 1998-2001, 31% for projects installed from 2002-2003, and 34% for projects installed in 2004- 2005. Once again, higher hub heights and larger rotor diameters (particularly relative to turbine nameplate capacity) are likely to be largely responsible for these increases in capacity factors.

Projects installed since 2005, however, have in general bucked this trend of rising capacity factors among newer projects: the capacity-weighted average 2009 capacity factors for projects built in 2006, 2007, and 2008 were 28%, 33%, and 29%, respectively. Though further analysis would be needed to fully assess the reasons for this leveling of capacity factors, potential explanations include:

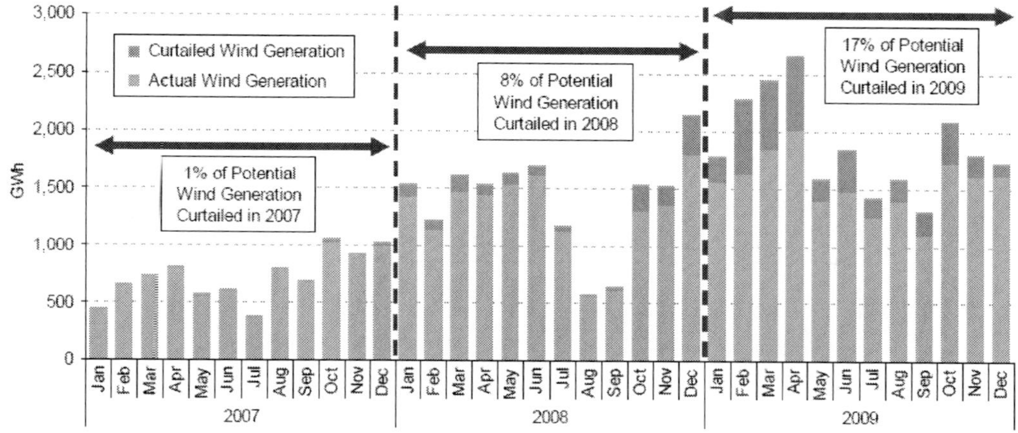

Source: ERCOT (2010)

Figure 32. History of Wind Energy Curtailment Within ERCOT: 2007-2009.

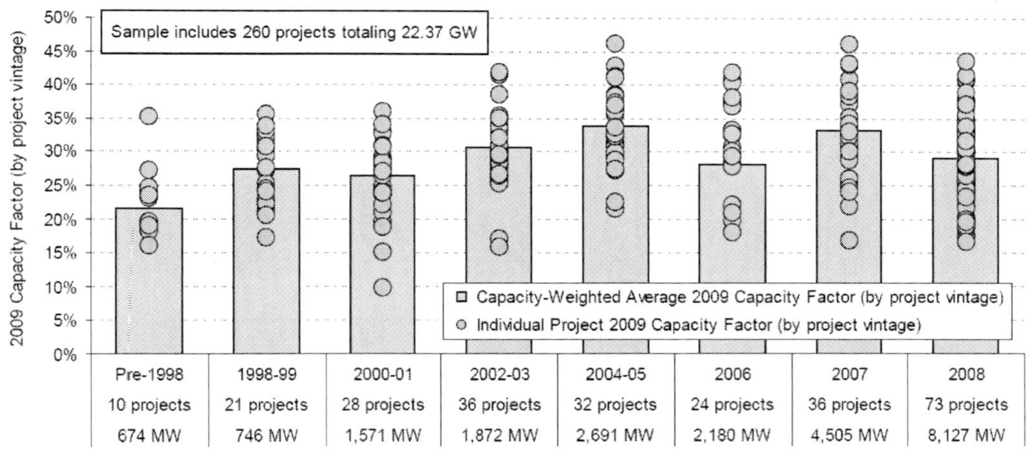

Source: Berkeley Lab

Figure 33. 2009 Project Capacity Factors by Commercial Operation Date.

Table 8. Capacity-Weighted Average 2009 Capacity Factors by Region and COD

Capacity Factor	Texas		East		Great Lakes		Northwest		New England		California		Heartland		Mountain		Hawaii	
Pre-1998	-		-		-		-		24.1%		21.4%		23.4%		-		-	
1998-99	-		-		19.0%		31.4%		-		30.7%		25.8%		32.2%		-	
2000-01	25.8%		22.7%		20.9%		23.6%		-		32.7%		32.3%		27.9%		-	
2002-03	25.3%		27.9%		17.1%		36.9%		-		30.4%		33.3%		30.7%		-	
2004-05	30.2%		29.9%		31.0%		32.8%		-		33.0%		36.7%		36.4%		-	
2006	21.6%		30.7%		-		28.6%		19.4%		30.2%		36.6%		34.9%		41.7%	
2007	30.2%		27.6%		32.6%		34.7%		32.9%		-		35.2%		35.4%		-	
2008	25.5%		27.6%		29.4%		26.1%		30.1%		38.9%		32.2%		32.8%		-	
Total	#	**26.2%**	#	**28.0%**	#	**30.2%**	#	**30.3%**	#	**31.1%**	#	**27.4%**	#	**33.2%**	#	**33.4%**	#	**41.7%**
Sample		**MW**		**MW**		**MW**		**MW**		**MW**		**MW**		**MW**		**MW**		**MW**
Pre-1998	0	0	0	0	0	0	0	0	1	6	8	642	1	26	0	0	0	0
1998-99	0	0	0	0	2	20	1	25	0	0	6	192	10	482	2	27	0	0
2000-01	4	669	6	83	1	30	3	373	0	0	1	67	9	227	4	123	0	0
2002-03	1	160	3	161	1	50	2	105	0	0	4	287	22	600	3	510	0	0
2004-05	4	461	2	349	1	54	5	434	0	0	3	130	15	1,062	2	200	0	0
2006	2	860	1	26	0	0	4	538	2	1	2	188	10	387	2	150	1	30
2007	7	1,135	4	169	5	679	4	654	2	44	0	0	10	1,049	4	776	0	0
2008	17	2,942	6	513	7	712	3	303	2	29	2	69	30	3,152	6	407	0	0
Total	**35**	**6,226**	**22**	**1,300**	**17**	**1,546**	**22**	**2,433**	**7**	**79**	**26**	**1,575**	**107**	**6,985**	**23**	**2,192**	**1**	**30**

Source: Berkeley Lab

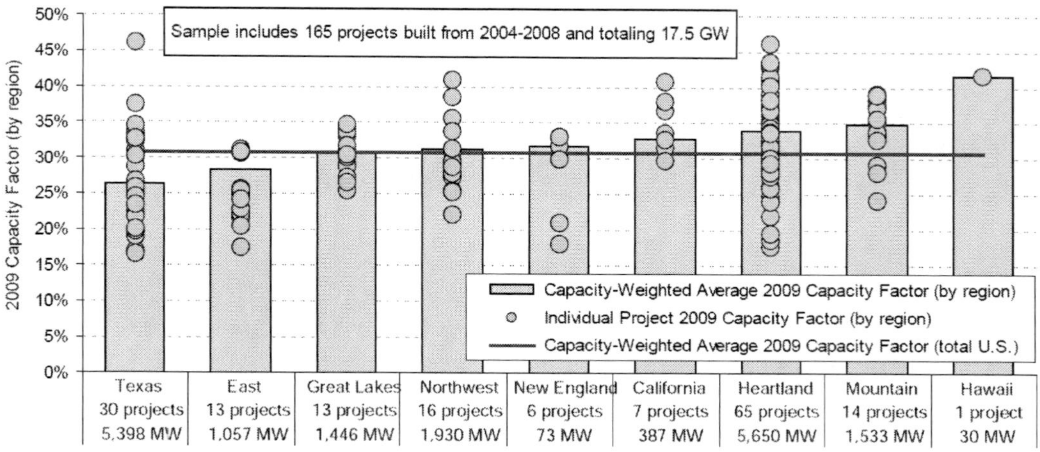

Source: Berkeley Lab

Figure 34. 2009 Project Capacity Factors by Region: 2004-2008 Projects Only.

- **Project Siting:** Developers may be reacting to increasing transmission constraints (or even just regionally differentiated wholesale electricity prices, or siting constraints) by focusing on those projects in their pipeline that may not be located in the best wind resource areas, but that do have ready access to unconstrained transmission (or higher-priced markets or readily available sites without long permitting times).
- **Technology Change:** Though increases in average turbine hub height and rotor diameter have been substantial, those increases have moderated in recent years (as discussed in an earlier section), yielding a weaker technical push towards higher capacity factors.
- **Turbine Reliability:** Some turbine manufacturers experienced blade and gearbox problems among their fleet of turbines installed in 2007 and 2008. Additionally, for the many projects completed in late 2008, the initial break-in period during which operational kinks are worked out may have extended well into 2009, negatively impacting 2009 capacity factors.

Trends in fleet-wide average capacity factors aside, the project-level spread shown in Figure 33 is enormous, with 2009 capacity factors ranging from 16.6% to 43.5% among projects built in the same year, 2008. Some of this spread is attributable to regional variations in wind resource quality. Figure 34 shows the regional variation in 2009 capacity factors, based on a sub-sample of wind power projects built from 2004 through 2008 (i.e., a period of relative stability in capacity factors, as shown in Figure 33). For this sample of projects, weighted-average capacity factors are the highest in Hawaii (above 40% on average) and the Mountain region (around 35% on average), and lowest in the East (below 30% on average) and in Texas (around 26% on average). The relatively low 2009 average capacity factor in Texas is largely caused by curtailment within ERCOT (see Figure 32): the ERCOT-wide 2009 capacity factor with curtailment of 25.8% would have been 3 1.1% were there no curtailment, an absolute difference of 5.2%. All other regions feature weighted-average capacity factors in 2009 that are in the 30- 35% range, which is similar to the national average among the overall 2004-2008 project sample. Given the small sample size in some regions, however, as well as the possibility that certain regions may have experienced a particularly good or bad wind resource year in 2009 or different levels of wind energy curtailment, care should be taken in extrapolating these results.

Though limited sample size is again a problem for many regions, Table 8 illustrates trends in 2009 capacity factors for projects with different commercial operation dates, by region.

Operations and Maintenance Costs Are Affected by the Age and Size of the Project, among Other Factors

Operations and maintenance (O&M) costs are a significant component of the overall cost of wind energy, but can vary substantially among projects. Market data on actual project-level O&M costs are not widely and readily available. Even where data are available, care must be taken in extrapolating historical O&M costs given the dramatic changes in wind turbine technology that have occurred over the last two decades, not least of which has been the up-scaling of turbine size (see Figures 14 and 15, earlier). Anecdotal evidence suggests that O&M costs and premature component failures continue to be key challenges to the wind power industry.

Berkeley Lab has compiled O&M cost data for 115 installed wind power projects in the United States, totaling 6,097 MW of capacity, with commercial operation dates of 1982 through 2008. These data cover facilities owned by both independent power producers and utilities, though data since 2004 are exclusively from utility-owned projects. A full time series of O&M cost data, by year, is available for only a small number of projects; in all other cases, O&M cost data are available for just a subset of years of project operations. Although the data sources do not all clearly define what items are included in O&M costs, in most cases the reported values appear to include the costs of wages and materials associated with operating and maintaining the facility, as well as rent (i.e., land lease payments). Other ongoing expenses, including taxes, property insurance, and workers' compensation insurance, are generally not included. Given the scarcity, limited content, and varying quality of the data, the results that follow may not fully depict the industry's challenges with O&M issues and expenditures; instead, these results should only be taken as illustrative of overall trends. Note also that the available data are presented in $/MWh terms, as if O&M represents a variable cost; in fact, O&M costs are in part variable and in part fixed. Although not presented here, expressing O&M costs in units of $/kW-year yields qualitatively similar results to those presented in this section.

Figure 35 shows project-level O&M costs by year of project installation (i.e., the most recent year that *original* equipment was installed, or the most recent year of project repowering). Here, O&M costs represent an average of annual project-level data available for the years 2000 through 2009. For example, for projects that reached commercial operations in 2008, only year 2009 data are available, and that is what is shown in the figure.[55] Many other projects only have data for a subset of years during the 2000-09 timeframe, either because they were installed after 2000 or because a full time series is not available, so each data point in the chart may represent a different averaging period over 2000-09. The chart highlights the 27 projects, totaling 2,450 MW, for which 2009 O&M cost data were available.

The data exhibit considerable spread, demonstrating that O&M costs are far from uniform across projects. However, Figure 35 suggests that projects installed more recently have, on average, incurred lower O&M costs. Specifically, capacity-weighted average 2000-09 O&M costs for projects in the sample constructed in the 1980s equal $32/MWh, dropping to $22/MWh for projects installed in the 1 990s, and to $9/MWh for projects installed in the 2000s. This drop in O&M costs may be due to a combination of at least two factors: (1) O&M costs generally increase as turbines age, component failures become more common, and manufacturer warranties expire[56]; and (2) projects installed more recently, with larger turbines and more sophisticated designs, may experience lower overall O&M costs on a per-MWh basis.

To help illustrate the possible influence of these two factors, Figure 36 shows annual O&M costs over time, based on the number of years since the last year of equipment installation. Annual data for projects of similar vintages are averaged together, and data for projects under 5 MW in size are excluded (to help control for the confounding influence of economies of scale). Note that, for each group, the number of projects used to compute the average annual values shown in the figure is limited, and varies substantially (from 4 to 18 data points per project-year for projects installed prior to 2000; from 6 to 15 data points per project-year for projects installed in 2000 through 2002; from 2 to 4 data points per project-year for projects installed in 2003 through 2005, and from 3 to 20 data points per project-year for projects installed in 2006 through 2008). With this limitation in mind, the figure shows that projects installed more recently have had, at least during their first two years of operation, lower O&M costs than those installed in earlier years.[57] In addition, pre-2000 projects show an upward trend in project-level O&M costs after the third year of project operation, though the sample size after year four is quite limited.

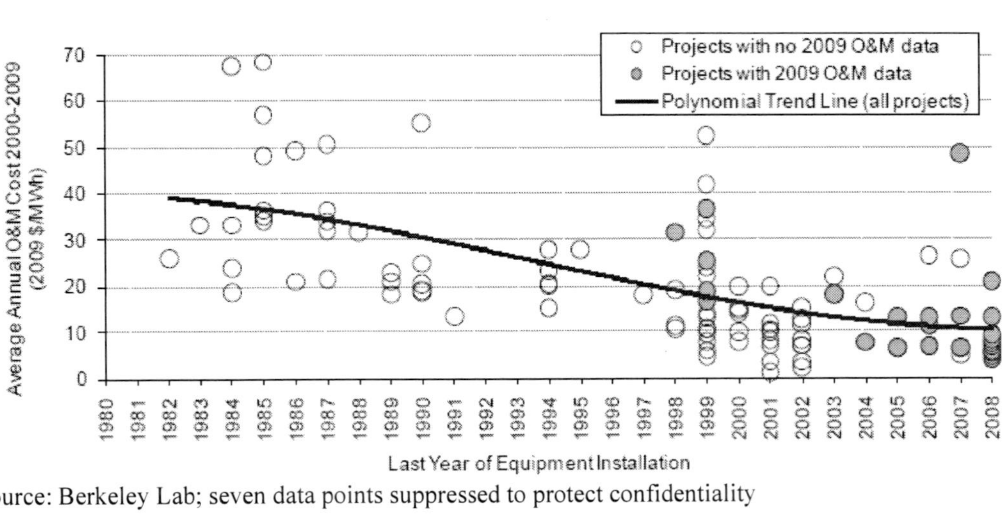

Source: Berkeley Lab; seven data points suppressed to protect confidentiality

Figure 35. Average O&M Costs for Available Data Years from 2000-2009, by Last Year of Equipment Installation.

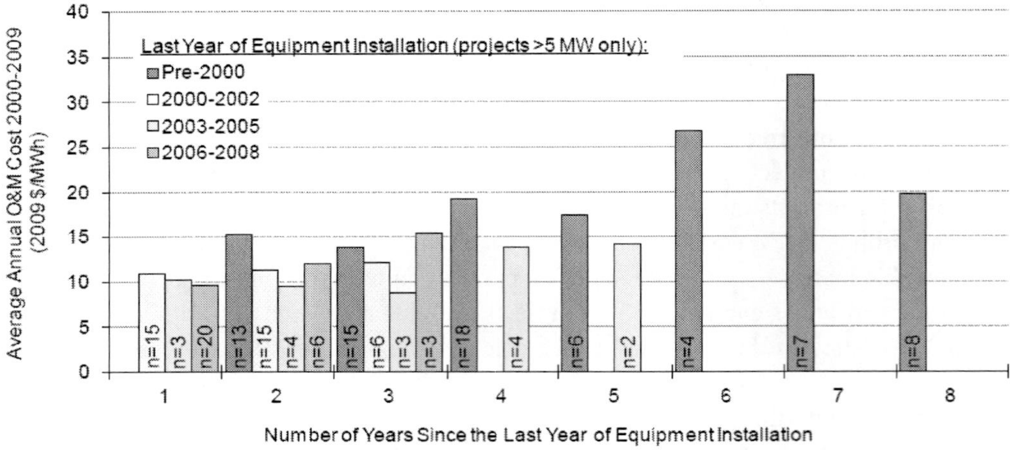

Source: Berkeley Lab; averages shown only for groups of two or more projects

Figure 36. Annual Average O&M Costs, by Project Age and Last Year of Equipment Installation.

Another variable that may impact O&M costs is project size. Figure 37 presents average O&M costs for 2000 through 2009 (as in Figure 35) relative to project size. Though the sample is too small for definite conclusions, project size does appear to have some impact on average O&M costs, with higher costs typically experienced by smaller projects.

5. Policy and Market Drivers

The Federal Policy Landscape Is Now More Favorable to Wind Energy than at Any Other Time in the Past Decade

A variety of policy drivers at both the federal and state levels have been important to the expansion of the wind power market in the United States At the federal level, perhaps the two most important policy incentives – at least in recent years – have been the PTC and accelerated tax depreciation. First established by the *Energy Policy Act of 1992*, the PTC provides a 10-year credit at a level that equaled 2.1 ¢/kWh in 2009 (adjusted annually for inflation), and increases to 2.2¢/kWh in 2010. The historical importance of the PTC to the U.S. wind power industry is illustrated by the pronounced lulls in wind power capacity additions in the three years (2000, 2002, and 2004) in which the PTC lapsed, as well as the increased development activity often seen during the year in which the PTC is otherwise scheduled to expire (see Figure 1). Accelerated tax depreciation, meanwhile, enables project owners to depreciate the vast majority of their investments over a five- to six-year period for tax purposes. An even-more-attractive 50% "bonus depreciation" schedule was in place during 2009, serving as an incremental driver of wind capacity additions in that year.

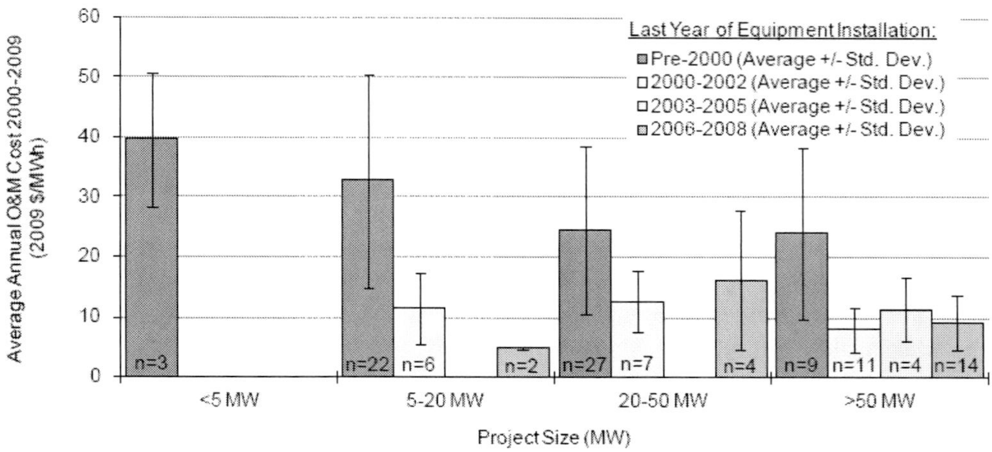

Source: Berkeley Lab; averages shown only for groups of two or more projects

Figure 37. Average O&M Costs for Available Data Years from 2000-2008, by Project Size.

Although these two federal incentives will likely remain important going forward (the PTC is currently in place through 2012, having been extended as part of the Recovery Act; while accelerated depreciation has no expiration date), in 2009 the PTC in particular was overshadowed by the Section 1603 Treasury cash grant program, enacted as part of the Recovery Act in February 2009. Acknowledging the conspicuous absence of tax equity investors in the market following the financial crisis of late 2008, Section 1603 of the Recovery Act enables wind power and other qualifying projects to elect a 30% cash grant in lieu of the PTC or ITC. Relative to the PTC, the 30% cash grant can provide a significant amount of value to wind power projects (Bolinger et al. 2010). Not surprisingly, then, the program has been heavily subscribed by wind power projects, which as of mid-July 2010 had received 88% of the more than $4.78 billion in Section 1603 cash grants awarded since the program's implementation in late-July 2009. More than 6,400 MW – i.e., more than 64% – of all new wind power capacity installed in the United States in 2009 chose the grant (Bolinger et al. 2010). The Section 1603 program should continue to be a strong driver for at least another year – wind power projects must begin construction by the end of 2010, and be operating by the end of 2012, in order to qualify for the grant – and several proposals exist to extend the program in some form.

Two other Recovery Act programs also played a (more limited) role in the wind energy sector in 2009. First, Title IV of the Recovery Act expanded a loan guarantee program that was originally enacted as part of the *Energy Policy Act of 2005*. The original program, under Section 1703, is targeted at projects that manufacture or utilize innovative clean energy technologies. The first two such guarantees for projects in the wind sector were conditionally awarded in 2009 and early 2010: Nordic Windpower received a $16 million loan guarantee to expand its two-bladed wind turbine manufacturing facility in Pocatello, Idaho, while developer First Wind received a $117 million loan guarantee for its 30 MW Kahuku wind project (which will include battery storage) in Hawaii. The Recovery Act also created a sister loan guarantee program, called the Section 1705 program, for projects using commercially proven technologies, but as of mid-July 2010, no wind power projects had yet been awarded Section 1705 loan guarantees.

Second, to encourage the growth of green manufacturing jobs in the United States, the Recovery Act created an advanced energy manufacturing tax credit, also known as the Section 48C credit, which provides a 30% tax credit for investments in new clean energy manufacturing facilities. More than 500 applications seeking in excess of $8 billion in Section 48C credits were submitted by the October 2009 deadline, exceeding the $2.3 billion program cap by more than a 3-to-1 margin. In early January 2010, 183 manufacturing projects spread across 43 states received credit allocations totaling $2.3 billion, with the wind energy sector capturing more than 10% of this total. Recipients are under no obligation to proceed with their projects, but in order to realize the credit, they must commission their facilities by February 2013. The Obama administration is seeking to extend the program for another year, with an additional $5 billion (Office of the Press Secretary, 2010).

In addition to the tax benefits and Recovery Act programs discussed above, a number of other federal policies have also helped to support different segments of the wind power industry in recent years. For example, because tax-exempt entities are unable to take direct advantage of tax incentives, the *Energy Policy Act of 2005* created the Clean Renewable Energy Bond (CREB) program, authorizing $800 million of what is effectively interest-free debt (though not without certain additional transaction costs) to eligible renewable projects.[58] Another $400 million of "old CREBs" were authorized in late 2006, followed by $2.4 billion in "new CREBs" authorized by the *Extension Act of 2008* ($800 million) and Recovery Act of 2009 ($1.6 billion). This old/new distinction is pertinent because "new CREBs" must follow a different set of rules – largely aimed at increasing the bonds' effectiveness – than existed under the "old CREBs." Applications for the $2.4 billion in "new CREBs" were due in early August 2009, and in late October, $2.2 billion in CREB allocations were awarded, with more than $450 million going to wind power projects.[59]

Finally, since 2003 the federal government has offered financial assistance to wind power (and other types of) projects that are located in rural areas. Specifically, Section 9006 of Title IX of *The Farm Security and Rural Investment Act of 2002* established *The Renewable Energy Systems and Energy Efficiency Improvements Program* (the Section 9006 program). Administered by the USDA, the Section 9006 program provided grants and loan guarantees to farmers, ranchers, and rural small businesses for assistance with purchasing renewable energy systems and making energy efficiency improvements. In May 2008, the Section 9006 program was converted to the *Rural Energy for America Program* (the REAP) by *The Food, Conservation, and Energy Act of 2008*. The REAP is little changed from the Section 9006 program – i.e., the REAP still targets agricultural producers and rural small businesses (including special purpose project companies set up specifically to own wind power projects) with grants and loan guarantees to encourage the installation of renewable energy systems and energy efficient upgrades. Grants are limited to the lesser of 25% of the project's cost or $500,000, whereas loan guarantees may not exceed $25 million (the combined amount of a grant and loan guarantee may not exceed 75% of a project's cost). In 2009, the USDA awarded more than $60 million through the REAP program (only a portion of which went to wind), while applications for an additional $100 million were due by June 30, 2010.

State Policies Play a Significant Role in Directing the Location and Amount of Wind Power Development

State policies continue to play a substantial role in directing the location and amount of wind power development. From 1999 through 2009, for example, 61% of the wind power capacity built in the United States was located in states with RPS policies; in 2009, this proportion was 57%. One new state (Kansas) established a mandatory RPS program in 2009, bringing the total to 29 states and Washington D.C. (see Figure 38); a number of additional states strengthened previously established RPS programs in 2009. In aggregate, existing state RPS policies would require roughly 73 GW of new renewable capacity by 2025, representing roughly 6% of total U.S. retail electricity sales in that year and 30% of projected load growth between 2000 and 2025.

Utility resource planning requirements in Western and Midwestern states have also helped spur wind power additions in recent years (especially as the prospect of future carbon regulations has been included as a variable in resource selection), as has growing voluntary customer demand for "green" power, especially among commercial customers. State renewable energy funds provide support for wind power projects (both financial and technical), as do a variety of state tax incentives. Finally, concerns about the possible impacts of global climate change are fueling interest by states and regions (as well as the federal government) to implement carbon reduction policies, a trend that is likely to increasingly underpin wind power expansion in the years ahead. The Northeast's Regional Greenhouse Gas Initiative cap-and-trade policy is now in operation, and carbon policies are also under discussion and being implemented in a number of other regions and states.

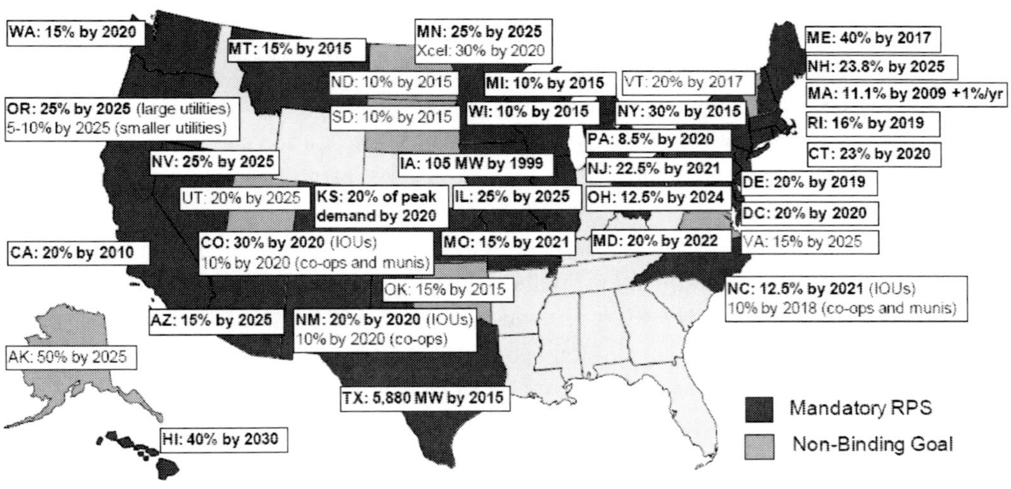

Source: Berkeley Lab

Figure 38. State RPS Policies and Non-Binding Renewable Energy Goals (as of July 2010).

Despite Progress on Overcoming Transmission Barriers, Constraints Remain

Transmission development appears to be gaining some traction. The North American Electric Reliability Corporation (NERC), for example, projects that transmission (100 kV and above) in the United States will increase by 31,400 circuit-miles, or about 8%, by 2018 (NERC 2009a), while the Brattle Group projects that annual transmission investment will exceed $10 billion going forward, compared to roughly $2 billion per year in the mid-1990s (Pfeifenberger et al. 2009).

Lack of transmission can be a barrier to wind power development. New transmission is particularly important for wind energy because wind power projects are constrained to areas with adequate wind speeds, which are often located at a distance from load centers. There is a mismatch between the relatively short timeframe needed to develop a wind power project compared to the time typically required to build new transmission. Uncertainty over siting and cost allocation, particularly for multi-state transmission lines, complicates transmission development.

With regards to transmission, several decisions at the federal level in 2009 created some concern among the wind power industry:

- The U.S. Court of Appeals ruled that FERC could use its backstop siting authority if a state withheld its decision for more than a year, but that it did not have the authority to override a state's decision to deny a transmission permit application (U.S. Fourth Circuit Court of Appeals 2009).
- The U.S. Court of Appeals remanded back to FERC the Commission's decision to approve spreading the costs of new transmission facilities above 500-kV in PJM to all transmission customers. Ruling that FERC had to better document how all transmission customers would benefit from the new transmission if all had to share in the cost, the Court's decision added some uncertainty to cost allocation policy (U.S. Seventh Circuit Court of Appeals 2009).
- FERC conditionally approved the Midwest ISO's petition to revise its pre-existing 50-50 cost share methodology to instead charge interconnecting generators 90% of the costs for transmission facilities rated 345 kV and above, and 100% for transmission facilities below 345 kV, though FERC also directed the Midwest ISO to file a revised cost allocation methodology by July 15, 2010 (FERC 2009). This policy contrasts with FERC's previous approval in 2007 of a California ISO proposal in which the cost of transmission for location- constrained resources would, initially, be covered in an ISO access charge (FERC 2007a).

Nonetheless, in June 2010, FERC issued a proposed transmission cost allocation rule that, among other things, would require that local and regional transmission planning processes incorporate the transmission needs that emanate from state or federal policies (such as RPS programs) and would establish principles that cost allocation proposals from grid providers must meet. The FERC proposal also indicates that, if agreement cannot be reached on cost allocation, FERC would itself develop a cost allocation method based on the record in that particular case (FERC 2010c).

States, grid operators, regional organizations, and DOE also continue to take proactive steps to encourage transmission investment to improve access to renewable resources. A non-exhaustive list of examples of these initiatives is presented below:

- **Texas Competitive Renewable Energy Zones (CREZ):** The Public Utility Commission of Texas (PUCT) encountered a slight delay with its $5 billion transmission plan for that state's CREZs. In January 2009, the PUCT awarded the development of CREZ transmission plan segments. The city of Garland challenged the decision, and in December 2009, the Court reversed and remanded the transmission construction allocation order (District Court 2009). The PUCT subsequently assigned the uncontested transmission segments, while creating a new regulatory docket for the reconsideration of the contested transmission segments.
- **Southwest Power Pool (SPP):** SPP received FERC approval for a new transmission cost allocation methodology in which costs will be paid by load and allocated between SPP and regions based on transmission voltage. Transmission projects of 300 kV and above will be funded 100% regionally across all of SPP, whereas costs for transmission projects between 100 kV and 300 kV will be allocated 1/3 regionally and 2/3 locally. Costs for transmission projects below 100 kV will be entirely recovered locally (SPP 2010). FERC's approval paves the way for SPP to construct $1.14 billion of "Priority Project" transmission in Kansas, Missouri, Oklahoma, and Texas.
- **Bonneville Power Administration (BPA) Network Open Season:** BPA's second annual network open season was held in June 2009, resulting in 83 transmission service requests for 4,867 MW, of which 2,599 MW were for wind power projects. This process subsequently led to signed transmission agreements for 1,553 MW, including 933 MW of wind power. BPA is also constructing the 500-kV McNary-John Day project, with $3.25 billion in increased borrowing authority from the Recovery Act (BPA 2009a); the line is expected to go into service in 2012 and could support 575 MW of new wind power capacity (BPA 2009b).
- **Western Area Power Administration (WAPA):** WAPA also received authority under the Recovery Act to increase its borrowing authority by up to $3.25 billion, and received over 200 transmission proposals to a WAPA-issued request. WAPA has so far made one loan of $161 million to Tonbridge Power for the Montana-Alberta Tie Transmission Line (WAPA 2009). Construction of the 230-kV line, primarily being developed to support wind power, is underway and is expected to be completed in 2011 (Tonbridge 2010).
- **Interconnection-Wide Transmission Planning:** In December 2009, DOE allocated $60 million in Recovery Act funds to promote collaborative long-term analysis and interconnection-wide transmission planning for the Eastern, Western, and Texas interconnections (DOE 2009).

A variety of efforts to proactively plan for transmission, often through analyses of state and regional renewable energy zones, also continued in 2009.[60] Finally, progress was made in 2009 on some of the transmission projects that are designed, in part, to support wind power, including:

- **California Tehachapi:** The first three segments of the Tehachapi transmission project were completed in May 2010. Meanwhile, the California PUC approved segments 4-11 in December 2009. Once fully operational, the Tehachapi transmission expansion is expected to be able to accommodate about 4,000 MW of wind power (SCE 2010).
- **Texas NextEra Transmission:** NextEra built a 200+mile 345-kV line to capture the higher wholesale prices that exist outside of West Texas, where most of wind power capacity is located. The line can transmit up to 950 MW and runs southeast from two of NextEra's wind power projects near Abilene (FPL 2009).
- **Maine Transmission:** In May 2010, the Maine PUC approved a $1.4 billion proposal for a 350-mile transmission line that could provide Maine wind power projects greater access to southern New England markets (CMP 2010).

Integrating Wind Energy into Power Systems Is Manageable, but Not Free of Costs, and Market Operators Are Implementing Methods to Accommodate Increased Penetration

During the past several years, there has been a considerable amount of attention paid to the potential impacts of wind energy on power systems. Concerns about, and solutions to, these issues have affected, and continue to impact, the pace of wind power deployment in the United States

Studies that have evaluated the operational impacts of wind energy on the power system have become increasingly sophisticated, resulting in a better accounting of the potential impacts and integration costs of increased wind energy penetration levels. Key trends among some of the more-recent studies include evaluating even higher levels of wind energy penetration, evaluating the integration of wind energy within larger electricity market areas, and identifying approaches to mitigate integration concerns. Additional recent high-level summaries and examples of wind energy integration in the United States and in other countries are available from IEEE (2009).

Two major studies of high penetrations of wind energy, each using different approaches and analysis tools, were completed in early 2010. The Eastern Wind Integration and Transmission Study (EWITS) examined land-based and offshore wind energy in the Eastern Interconnection at penetrations of up to 30% on an energy basis (EnerNex 2010). The Western Wind and Solar Integration Study (WWSIS) examined wind and solar energy in the Western Interconnection with a particular focus on the WestConnect footprint; the highest penetration examined in the WWSIS was a scenario with 30% wind and 5% solar on an energy basis within the WestConnect footprint and 20% wind and 3% solar energy in the rest of the Western Interconnection (GE 2010).[61] Both studies found that, with significant improvements in operational practices, it is technically feasible to operate the power system with high penetrations of wind energy. Changes in operational practices that were found to be beneficial include increased procurement of operating reserves, greater use of sub-hourly generation scheduling, enhanced flexibility and cycling of natural gas and coal plants, incorporation of state-of-the-art wind forecasting into system operations, utilization of demand response, and increased cooperation or consolidation of balancing areas. Although the level of detail in the transmission analysis differed between the two studies, both involved extensive transmission expansion to deliver wind energy resources to load centers and to manage the additional variability and uncertainty.

In addition to these two studies at the interconnection level, Table 9 provides a selective listing of results from wind energy integration cost studies[62,63] completed from 2003 through early 2010.[64] Similar information is presented in Figure 39 at various levels of wind power *capacity* penetration. Because methods vary and a consistent set of operational impacts has not been included in each study, results from the different analyses are not fully comparable. Note also that the rigor with which the various studies have been conducted varies, as does the degree of peer review. Nonetheless, key conclusions that continue to emerge from the growing body of integration literature include the following:

- Wind energy integration costs are below $10/MWh – and often below $5/MWh – for wind power capacity penetrations up to or exceeding 40% of the peak load of the system in which the wind power is delivered. [65] Variations in estimated costs across studies are due, in part, to differences in methodologies, definitions of integration costs, power system and market characteristics, wind energy penetration levels, and fuel price assumptions.
- Regulation impacts are often found to be relatively small, whereas the impacts of wind energy on load-following and unit commitment are typically found to be more significant.
- Larger balancing areas, such as those found in RTOs and ISOs, make it possible to integrate wind energy more easily and at lower cost than is the case in smaller balancing areas.
- The successful use of wind power forecasts by system operators can significantly reduce integration challenges and costs. Wind forecasts are most accurate and effective when aggregated across large, electrically interconnected areas.

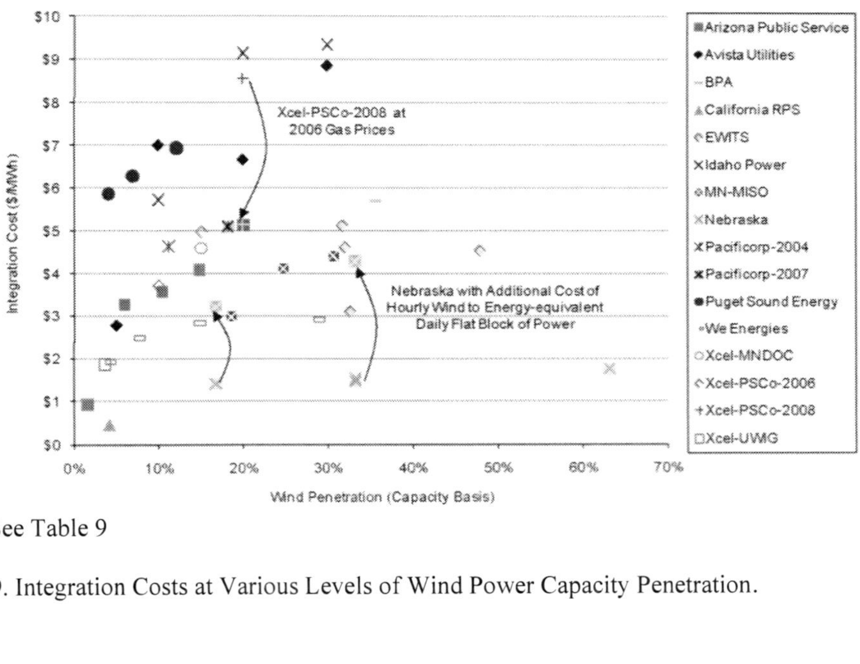

Source: See Table 9

Figure 39. Integration Costs at Various Levels of Wind Power Capacity Penetration.

- Intra-hour scheduling (e.g., 5-10 minute schedules) provides access to flexibility in conventional power plants that lowers the costs of integrating wind energy.
- Wind energy integration costs tend to rise with increasing natural gas prices, though the economic value of wind energy also increases with higher gas prices.

Table 9. Key Results from Selected Wind Energy Integration Cost Studies[66]

			Integration Cost ($/MWh)				
Year	Study	Wind Capacity Penetration	Regulation	Load Following	Unit Commit.	Gas Supply	TOTAL
2003	Xcel-UWIG	3.5%	0	0.41	1.44	-	1.85
2003	We Energies	29%	1.02	0.15	1.75	-	2.92
2004	Xcel-MNDOC	15%	0.23	-	4.37	-	4.60
2005	PacifiCorp-2004	11%	0	1.48	3.16	-	4.64
2006	Calif. (multi-year)*	4%	0.45	trace	trace	-	0.45
2006	Xcel-PSCo	15%	0.20	-	3.32	1.45	4.97
2006	MN-MISO**	31%	-	-	-	-	4.41
2007	Puget Sound Energy	12%	-	-	-	-	6.94
2007	Arizona Pub. Service	15%	0.37	2.65	1.06	-	4.08
2007	Avista Utilities	30%	1.43	4.40	3.00	-	8.84
2007	Idaho Power	20%	-	-	-	-	7.92
2007	PacifiCorp-2007	18%	-	1.10	4.00	-	5.10
2008	Xcel-PSCo***	20%	-	-	-	-	8.56
2009	Bonneville (BPA)+	36%	0.22	1.14	-	-	5.70
2010	EWITS++	48%	-	-	1.61	-	4.54
2010	Nebraska+++	63%	-	-	-	-	1.75

* Regulation costs represent 3-year average.

** Highest over 3-year evaluation period.

*** This integration cost reflects a $10/MMBtu natural gas price scenario. This cost is much higher than the integration cost calculated for Xcel-PSCo in 2006, in large measure due to the higher natural gas price: had the gas price from the 2006 study been used in the 2008 study, the integration cost would drop to $5.13/MWh.

+ Costs in $/MWh assume 31% capacity factor. Aside from regulation and following reserves, the costs of BPA's imbalance reserves are $4.33/MWh.

++ The unit commitment costs listed in EWITS are the cost of day-ahead wind forecast error; the remaining integration costs included in the total are for shorter term variable reserves that account for regulation and short-term forecast errors (energy imbalance).

+++ These integration costs only capture regulating reserves and day-ahead forecast error. A sensitivity case in this study shows that integration costs increase if the differences between the actual hourly deliveries of wind energy are compared to daily flat block of power. The increased costs are shown in Figure 39.

Sources: Brooks et al. (2003) [Xcel-UWIG]; Electrotek Concepts, Inc. (2003) [We Energies]; EnerNex Corp. and Wind Logics, Inc. (2004) [Xcel-MNDOC]; PacifiCorp (2005) [Pacificorp-2004]; Shiu et al. (2006) [Calif. (multiyear)]; EnerNex Corp. (2006) [Xcel-PSCo]; EnerNex Corp. and Windlogics Inc. (2006) [MN-MISO]; Puget Sound Energy (2007) [Puget Sound Energy]; Acker

(2007) [Arizona Pub. Service]; EnerNex Corp. (2007) [A vista Utilities]; EnerNex Corp. and Idaho Power Co. (2007) [Idaho Power]; PacifiCorp (2007) [PacifiCorp-2007]; EnerNex Corp. (2008) [Xcel-PSCo]; BPA (2009) [Bonneville]; EnerNex Corp (2010) [EWITS]; EnerNex et al. (2010) [Nebraska]

Many ISOs and utilities are also continuing to take important steps to mitigate the challenges faced with integrating larger quantities of wind energy. Centralized wind forecasting systems are currently in place at the PJM, Electric Reliability Council of Texas, Midwest ISO, New York ISO, California ISO, Southern California Edison, and Xcel Energy; while the BPA is currently developing wind forecasting systems (Porter and Rogers 2010). Northern Tier Transmission Group, Columbia Grid, and WestConnect, meanwhile, are jointly investigating projects that will increase power system flexibility, including the creation of a dynamic scheduling communications infrastructure, sharing of area control errors, and intra-hour scheduling and balancing. These initiatives have broad benefits, including better utilization of the transmission system and providing increased flexibility to integrate wind energy.

Some utilities are now directly charging wind power projects for balancing services or are reducing posted 'avoided cost' contract price payments to account for the costs of integrating wind energy. BPA, for example, includes a wind energy balancing charge in its transmission tariff equivalent to about $5 .70/MWh. FERC conditionally approved a higher generator regulation and frequency response services charge for wind energy in the Westar Energy balancing area equivalent to about $0. 80/MWh; this tariff is still in FERC proceedings and may be revised (FERC 2010a). Idaho Power, Avista, and PacifiCorp, meanwhile, all discount their avoided cost payments for qualifying wind power projects by an integration rate that ranges from 7-9% of the avoided cost rate, up to $6.50/MWh (IPUC 2010).

At a national level, NERC and FERC have also been focused on identifying methods to reliably and economically integrate wind energy into the bulk power system. NERC released a comprehensive report in 2009 with several recommendations for changing planning and operational procedures to maintain reliable operation with high levels of variable generation (NERC 2009b). Following the report, NERC outlined a three-year work plan to further develop and improve standards and practices for integrating variable generation. FERC, meanwhile, issued a notice of inquiry seeking public comment on whether to reform any of its rules or procedures to ensure that increased amounts of variable energy resources can be accommodated with just and reasonable rates and without undue discrimination (FERC 201 0b). Over one hundred reply comments have been filed by parties throughout the United States proposing potential steps to better facilitate the integration of variable generation.

6. FUTURE OUTLOOK

Despite the growth in installed capacity in 2009, the combination of the financial crisis, lower wholesale electricity prices, and lower demand for renewable energy has taken a toll on the wind power industry. During 2009, the relative economic position of wind energy became more challenging, orders for new turbines slowed, and turbine and component manufacturers announced some layoffs. Wind power capacity additions in 2009 were buoyed, in part, by projects that were initially slated to be completed in 2008 but that carried over into 2009 when the PTC was extended, somewhat masking the underlying challenges facing the sector. With federal incentives now extended through 2012, there is less motivation to complete projects in 2010 (though many projects will likely start construction in 2010 in order to qualify for the 30% Treasury cash grant). As a result, though the Recovery Act has helped to alleviate financing challenges, expectations of a slower year in 2010 remain.

A variety of forecasts suggest that wind power installations in 2010 may fall within the range of 5,500 MW to 8,000 MW, a drop of 20-45% compared to the nearly 10,000 MW installed in 2009 (see Table 10). This contraction is reflected in results for the first half of 2010, in which just 1,240 MW of wind power were installed – i.e., 57% less than the amount installed in the first half of 2008, and 71% below the pace set in the first half of 2009 (AWEA 2010c). After a slower year in 2010, these predictions show market resurgence in 2011 and 2012, with annual installations ranging from 8,100 to 15,000 MW depending on the forecast and year. From 2010 through 2012, these forecasts predict cumulative wind power additions of 24 to 33 GW; this amount of new wind power capacity would provide roughly 30-40% of EIA's projected growth in total U.S. electricity demand over the 2010-2012 timeframe. Though not shown in Table 10, any projections beyond 2012 are rendered considerably less certain by the scheduled expiration of a number of policies at the end of that year, including the PTC, the ability to elect a 30% ITC in lieu of the PTC, and the ability to receive the 30% cash grant for projects that initiated construction by the end of 2010.

Notwithstanding the anticipated slowdown in 2010, these growth projections would likely ensure that the United States retains its 2009 position as the second-largest wind energy market in the world in terms of annual capacity additions. Driven by rapidly growing energy demands and strong policy support, China is widely expected to lead to the world in annual wind power capacity additions in the coming years. Industrial policy and market conditions have also resulted in the growing dominance of Chinese wind turbine manufacturers within the Chinese market, and those manufacturers are beginning to explore export strategies. U.S. manufacturing of turbines and components is also expected to continue to grow, as already-announced manufacturing facilities come on line, as existing facilities reach full capacity and expand, and as new announcements and investments are made. In part as a result, and in a continuation of recent trends, the historically-dominant wind turbine suppliers in the United States market are likely to face growing competition from new entrants over this timeframe.

Uncertainties about market performance even over the 2010-2012 timeframe are the result of underlying uncertainties about market and policy drivers. On the positive side, the wind power industry now has stronger federal policy support than at any time in the last decade, and state policies have become more aggressive. Additionally there are prospects for further federal policy support through some combination of a continuation of (or variants to) the Treasury Grant program, federal RPS legislation, climate legislation, and policies intended to spur new transmission investments, as well as continued state renewable energy and climate policy initiatives. With wind turbine prices now dropping, the trend of increasing project-level costs and prices experienced over the last several years is also expected to slow and even reverse, improving the comparative economic position of wind.

Table 10. Forecasts for Annual U.S. Wind Capacity Additions (MW)

Source	2010	2011	2012	Cumulative Additions 2010-2012
EIA	7,310	10,200	10,330	27,840
BTM	8,000	10,000	15,000	33,000
IHS EER	7,130	9,830	9,340	26,300
Bloomberg NEF	7,390	8,535	8,610	24,535
Macquarie	7,500	8,100	8,700	24,300
UBS	6,950	9,380	10,780	27,110
AWEA	5,500-7,500	--	--	--

Source: Bloomberg NEF (2010b), BTM (2010), EIA (2010), IHS Emerging Energy Research (2010), Macquarie (2010), UBS (2010), AWEA (2010c)

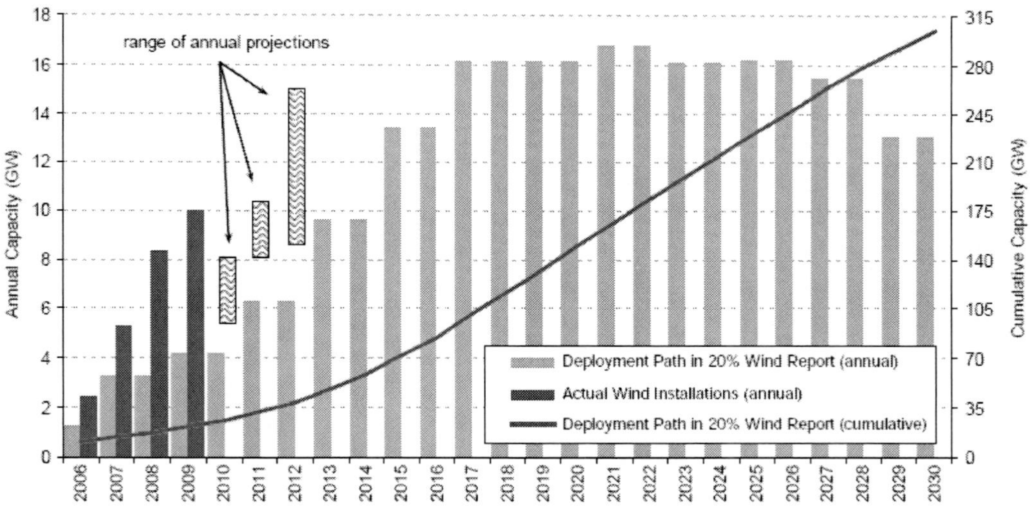

Source: DOE (20% wind scenario); AWEA (historical additions); Table 10 (projected additions)

Figure 40. Wind Power Capacity Growth: 20% Wind Report, Actual Installations, Projected Growth.

On the other hand, with the window of eligibility for the Treasury Grant program scheduled to close at the end of 2010 and the tax equity market for wind energy not fully recovered, near-term growth may be hampered. Natural gas prices and near-term price expectations have plummeted, making wind energy's primary competitor more economically attractive than in recent years. And, with the much-lower wholesale electricity price environment, merchant wind power development – which had grown dramatically in recent years – may slow. The significant wind energy growth in recent years has also exceeded aggregate state RPS demands, resulting in softer demand from state RPS markets in the near term. Wind power additions are increasingly constrained by inadequate transmission infrastructure, and while progress is being made to alleviate those constraints, the build-out of transmission infrastructure will take time. Finally, in California and the Southwest in particular, wind energy is beginning to face stiff competition with solar energy in meeting state renewable energy requirements.

Regardless of these competing trends, wind power capacity additions over the past several years, and those projected from 2010-2012, put the United States on a trajectory that may lead to 20% of the nation's electricity demand coming from wind energy by 2030. In May 2008, the U.S. Department of Energy, in collaboration with its national laboratories, the wind power industry, and others, published a report that analyzed the technical and economic feasibility of achieving 20% wind energy penetration by 2030 (DOE 2008). In addition to finding no insurmountable barriers to reaching 20% wind energy penetration, the report also laid out a potential wind power deployment path that started at 3.3 GW/year in 2007, increasing to 4.2 GW/year by 2009, 6.4 GW/year by 2011, 9.6 GW/year by 2013, 13.4 GW/year by 2015, and roughly 16 GW/year by 2017 and thereafter, yielding cumulative wind power capacity of 305 GW by 2030. Historical growth over the last four years puts the United States on a trajectory exceeding this deployment path, and the projected growth presented in Table 10 would ensure that the United States remains in that position through 2012 (Figure 40).

Ramping up to an annual installation rate of 16 GW per year, and maintaining that rate for a decade, is, however, far from pre-determined. The record 10 GW installation pace in 2009 was achieved, in part, because of the previously-pending end-of-2008 expiration of the PTC. Moreover, federal policy towards wind energy remains uncertain after 2012. Whether the roughly 16 GW per year pace needed for wind power to contribute 20% of the nation's electricity by 2030 can be achieved and maintained remains to be seen.

In addition to stable, long-term promotional policies, the DOE (2008) report suggests four other areas where supportive actions may be needed in order to reach such annual installation rates. First, the nation will need to invest in significant amounts of new transmission infrastructure designed to access remote wind resources. Second, to more-effectively integrate wind power into electricity markets, larger power control regions, better wind forecasting, and increased investment in fast-responding generating plants will be required. Third, streamlined siting and permitting procedures will need to be established to allow wind power developers to identify appropriate project locations and move from wind resource prospecting to construction quickly. Finally, enhanced research and development efforts in both the public and private sector will be required to lower the cost of offshore wind power, and incrementally improve conventional land- based wind energy technology.

APPENDIX. SOURCES OF DATA PRESENTED IN THIS REPORT

Installation Trends

Data on wind power additions in the United States come from AWEA, though methodological differences noted throughout this report result in some difference in the data presented here relative to AWEA (2010a). Annual wind power capital investment estimates derive from multiplying these wind power capacity data by weighted-average capital cost data, provided elsewhere in the report. Data on non-wind electric capacity additions come primarily from EIA (for years prior to 2009) and Ventyx's Velocity database (for 2009), except that solar data come from the Interstate Renewable Energy Council (IREC) and Berkeley Lab. Data on the small wind turbine market come from AWEA (20 10b).

Information on offshore wind power development activity in the United States was compiled by Berkeley Lab, NREL, and Energetics.

Global cumulative (and 2009 annual) wind power capacity data come from BTM (2010), but are revised to include the U.S. wind power capacity used in the present report. Historical cumulative and annual worldwide capacity data come from BTM Consult and the Earth Policy Institute. Wind energy as a percentage of country-specific electricity consumption is based on end-of-2009 (and end-of-2006/07/08) wind power capacity data and country-specific assumed capacity factors that primarily come from BTM (2010). For the United States, the performance data presented in this report are used to estimate wind energy production. Country-specific projected wind generation is then divided by projected electricity consumption in 2010 (and 2007/08/09), based on actual 2007 consumption and a country-specific growth rate assumed to be the same as the rate of growth from 2002 through 2007 (these data come from EIA).

The wind power project installation map was created by NREL, based in part on AWEA's database of projects and in part on data from Platts on the location of individual projects. Effort was taken to reconcile the AWEA project database and the Platts-provided project locations, though some discrepancies remain. Wind energy as a percentage contribution to statewide electricity generation is based on AWEA installed capacity data for the end of 2009 and the underlying wind power project performance data presented in this report. Where necessary, judgment was used to estimate state-specific capacity factors. The resulting state wind generation is then divided by in-state total electricity generation in 2009, based on EIA data.

The listing of wind power capacity serving specific electric utilities comes from AWEA's *U.S. Wind Industry Annual Market Report* (AWEA 201 0a), with two exceptions: (1) the Empire District Electric Company was added to AWEA's "top twenty" investor-owned utility list at position number 14; and (2) Minnkota Power Cooperative's wind capacity was corrected to 357 MW (AWEA (2010a) shows 290 MW). To translate this capacity to projected utility-specific annual electricity generation, regionally appropriate wind power capacity factors are used. The resulting utility-specific projected wind generation is then divided by the aggregate national retail sales of each utility in 2008 (based on EIA Form-861 data). Only utilities with 100 MW or more of wind power capacity are included in these calculations. In the case of G&T cooperatives and power authorities that provide power to other cooperatives and municipal utilities (but do not directly serve load themselves), this report uses 2008 retail sales from the electric utilities served by those G&T cooperatives and power authorities. In some cases, these individual utilities may be buying additional wind directly from other projects, or may be served by other G&T cooperatives or power authorities that supply wind. In these cases, the penetration percentages shown in the report may be somewhat misleading.

Data on wind power capacity in various interconnection queues come from a review of publicly available data provided by each ISO, RTO, or utility. Only projects that were active in the queue at the end of 2009, but that had not yet been built, are included. Suspended projects are not included in these listings. Data on projects that are in the nearer-term development pipeline come from Ventyx (2010).

Industry Trends

Turbine manufacturer market share, average turbine size, and average project size are derived from the AWEA wind power project database, with some processing by Berkeley Lab. Information on turbine hub heights and rotor diameters were compiled by Berkeley Lab based on information provided by turbine manufacturers, standard turbine specifications, FAA data, web searches, and other sources.

Information on wind turbine and component manufacturing come from NREL, AWEA, and Berkeley Lab, based on a review of press reports, personal communications, and other sources. The listings of manufacturing and supply chain facilities are not intended to be exhaustive. Data on aggregate U.S. imports and exports of wind power equipment come primarily from the U.S. International Trade Commission, and can be obtained from the USITC's DataWeb (http://dataweb.usitc.gov/).

Information on wind power developer consolidation and financing trends were compiled by Berkeley Lab. Wind project ownership and power purchaser trends are based on a Berkeley Lab analysis of the AWEA project database.

Price, Cost, and Performance Trends

Wind power price data are based on multiple sources, including prices reported in FERC's *Electronic Quarterly Reports*, FERC Form 1, avoided cost data filed by utilities, pre-offering research conducted by bond rating agencies, and a Berkeley Lab collection of power purchase agreements. Wholesale electricity price data were compiled by Berkeley Lab from the IntercontinentalExchange (ICE) as well as Ventyx' s Velocity database (which itself derives wholesale price data from the ICE and the various ISOs). Earlier years' wholesale electricity price data come from FERC (2007b, 2005). REC price data were compiled by Berkeley Lab based on information provided by Evolution Markets and Spectron.

Berkeley Lab used a variety of public and some private sources of data to compile capital cost data for a large number of U.S. wind power projects. Data sources range from pre-installation corporate press releases to verified post-construction cost data. Specific sources of data include: EIA Form 412, FERC Form 1, various Securities and Exchange Commission filings, various filings with state public utilities commissions, *Windpower Monthly* magazine, AWEA's *Wind Energy Weekly*, DOE/EPRI's Turbine Verification Program, *Project Finance* magazine, various analytic case studies, and general web searches for news stories, presentations, or information from project developers. For 2009 projects, data from the Section 1603 Treasury Grant program were used extensively. Some data points are suppressed in the figures to protect data confidentiality. Because the data sources are not equally credible, little emphasis should be placed on individual project-level data; instead, it is the trends in those underlying data that offer insight. Only wind power cost data from the contiguous lower-48 states are included.

Wind turbine transaction prices were compiled by Berkeley Lab. Sources of transaction price data vary, but most derive from press releases and press reports. In part because wind turbine transactions vary in the services offered, a good deal of intra-year variability in the cost data is apparent. Additionally, the data do not adequately capture the rumored softening of the wind turbine market since late 2008, as relative few publicly reported wind turbine sales transactions exist since that time.

Wind power project performance data are compiled overwhelmingly from two main sources: FERC's *Electronic Quarterly Reports* and EIA Form 923. Additional data come from FERC Form 1 filings and, in several instances, other sources. Where discrepancies exist among the data sources, those discrepancies are handled based on the judgment of Berkeley Lab staff. Data on curtailment in Texas are from ERCOT and in the Midwest from MISO.

Wind project operations and maintenance costs come primarily from two sources: EIA Form 412 data from 2001-2003 for private power projects and projects owned by POUs, and FERC Form 1 data for IOU-owned projects. Some data points are suppressed in the figures to protect data confidentiality.

Policy and Market Drivers

The wind energy integration, transmission, and policy sections were written by staff at Berkeley Lab and Exeter Associates, based on publicly available information.

Future Outlook

This section was written by staff at Berkeley Lab, based largely on publicly available information.

REFERENCES

Acker, T. (2007). *Arizona Public Service Wind Integration Cost Impact Study*. Prepared for Arizona Public Service Company. Flagstaff, Arizona: Northern Arizona University.

American Wind Energy Association (AWEA). (2010a). *U.S. Wind Industry Annual Market Report: Year Ending 2009*. Washington, DC: American Wind Energy Association.

American Wind Energy Association (AWEA). (2010b). *AWEA Small Wind Turbine Global Market Survey, Year Ending 2009*. Washington, DC: American Wind Energy Association.

American Wind Energy Association (AWEA). (2010c). *A WEA Mid-Year 2010 Market Report*. Washington, DC: American Wind Energy Association.

Bailey, D. (2010). "Wind developers shun US federal loan programme." *Windpower Monthly. 26(4)*, 32-34.

Bloomberg New Energy Finance. 2010a. *Q1 Wind Market Outlook*. February 2010. Bloomberg New Energy Finance.

Bloomberg New Energy Finance. (2010b). *Q2 Wind Market Outlook*. June 2010. Bloomberg New Energy Finance.

Bolinger, M., Wiser, R. & Darghouth, N. (2010). *Preliminary Evaluation of the Impact of the Section 1603 Treasury Grant Program on Renewable Energy Deployment in 2009.* LBNL-3 1 88E. Berkeley, California: Lawrence Berkeley National Laboratory.

BPA. (2009a). *Summary of Eligible TSRs for 2009 Network Open Season*. July 22, 2009. http://www.transmission mmary _07-22-09.pdf.

BPA. (2009b). *McNary-John Day 500-kilovolt Transmission Line*. September 9, 2009. http://www.bpa.gov/corporate/RecoveryAct/mcnary-johnday.cfm.

Brooks, D., Lo, E., Zavadil, R., Santoso, S. & Smith. J. (2003). *Characterizing the Impact of Significant Wind Generation Facilities on Bulk Power System Operations Planning: Xcel Energy – North Case Study*. Prepared for the Utility Wind Integration Group. Arlington, Virginia: Electrotek Concepts.

BTM Consult. (2010). *International Wind Energy Development: World Market Update 2009.* Ringkobing, Denmark: BTM Consult ApS.

California ISO (CAISO). (2007). *Integration of Renewable Resources*. Folsom, California: California Independent System Operator.

Chadbourne & Parke. (2010a). "Update: Tax Equity Market." *Project Finance Newswire*. April 8-20.

Chadbourne & Parke. (2010b). "The Search for Lowest Cost Capital." *Project Finance Newswire*. July, 1-9.

Charles River Associates. (2010). *SPP WITF Wind Integration Study*. Little Rock, Arkansas: Southwest Power Pool.

CMP. (2010). *Maine Utility Regulators Approve $1.4 Billion Transmission Line Investment.* May 14, 2010. http://www.cmpco.com/OurCompany/ News/2010/ news100514.html.

David, A. (2009). *Wind Turbines: Industry and Trade Summary*. ITX-02. Washington, DC: U.S. International Trade Commission.

David, A. (2010). *Impact of Wind Energy Installations on Domestic Manufacturing and Trade.* ID-25. Washington, DC: U.S. International Trade Commission.

Department of Energy (DOE). (2008). *20% Wind Energy by 2030: Increasing Wind Energy's Contribution to U.S. Electricity Supply*. DOE/GO-102008-2567. Washington, D.C.: U.S. Department of Energy.

Department of Energy (DOE). (2009). *Secretary Chu Announces Efforts to Strengthen U.S. Electric Transmission Networks*. December 18, 2009. http://www.energy.

District Court. (2009). *City of Garland vs. Public Utility Commission of Texas*. January 15, 2010.

Electrotek Concepts, Inc. (2003). *Systems Operations Impacts of Wind Generation Integration Study*. Prepared for We Energies. Knoxville, Tennessee: Electrotek Concepts.

Energy Information Administration (EIA). (2010). *Annual Energy Outlook 2010*. DOE/EIA-03 83(2010). Washington D.C.: Energy Information Administration.

EnerNex Corp. (2006). *Wind Integration Study for Public Service Company of Colorado*. Prepared for Xcel Energy. Denver, Colorado.

EnerNex Corp. (2007). *Final Report A vista Corporation Wind Integration Study*. Knoxville, Tennessee: EnerNex Corporation.

EnerNex Corp. (2008. *Wind Integration Study for Public Service of Colorado, Addendum, Detailed Analysis of 20% Wind Penetration*. Prepared for Xcel Energy. Denver, Colorado: Xcel Energy.)

EnerNex Corp. (2010. *Eastern Wind Integration and Transmission Study.* NREL/SR-550-47078. Golden, Colorado: National Renewable Energy Laboratory.)

EnerNex Corp. and Idaho Power Co. (2007). *Operational Impacts of Integrating Wind Generation into Idaho Power's Existing Resource Portfolio: Report Addendum*. Boise,

Idaho: Idaho Power Company.

EnerNex Corp., Ventyx, and Nebraska Power Association. (2010). *Nebraska Statewide Wind Integration Study*. NREL/ SR-550-47519. Golden, Colorado: National Renewable Energy Laboratory.

EnerNex Corp. and Wind Logics, Inc. (2004). *Wind Integration Study—Final Report.* Prepared for Xcel Energy and Minnesota Department of Commerce. Knoxville, Tennessee: EnerNex Corporation.

EnerNex Corp. and Windlogics Inc. (2006). *Final Report – 2006 Minnesota Wind Integration Study, Volume I.* Prepared for the Minnesota Public Utilities Commission. Knoxville, Tennessee: EnerNex Corporation.

ERCOT. (2010). *Personal communication with Dan Jones, ER COT Independent Market Monitor.* February 2010.

European Wind Energy Association (EWEA). (2010). *Wind in Power: 2009 European Statistics*. Brussels, Belgium: European Wind Energy Association.

Federal Energy Regulatory Commission (FERC). (2005). *2004 State of the Markets Report.* Washington, DC: Federal Energy Regulatory Commission.

Federal Energy Regulatory Commission (FERC). (2007a). *Order Granting Petition for Declaratory Order*. 119 FERC ¶ 61,061, April 19, 2007.

Federal Energy Regulatory Commission (FERC). (2007b). *2006 State of the Markets Report.* Washington, DC: Federal Energy Regulatory Commission.

Federal Energy Regulatory Commission (FERC). (2009). *Midwest Independent Transmission System Operator, Inc. and The Midwest ISO Transmission Owners*. 129 FERC ¶ 61,060.

Federal Energy Regulatory Commission (FERC). (2010a). *Order conditionally accepting proposed tariff revisions re Westar Energy, Inc. under ER09-1273.* 130 FERC ¶ 61,215.

Federal Energy Regulatory Commission (FERC). (2010b). *Notice of Inquiry Seeking Public Comment on the Integration of Variable Energy Resources under RM10-11*. Washington D.C.: Federal Energy Regulatory Commission.

Federal Energy Regulatory Commission (FERC). (2010c). *Transmission Planning and Cost Allocation by Transmission Owning and Operating Public Utilities*. 131 FERC ¶ 61,253.

FPL. (2009). *FPL Group, Inc. Q3 2009 Earnings Call Transcript*. October 27, 2009. http://seekingalpha.com/article/169267-fpl-group-inc-q3-2009-earnings- 1.

GE Energy. (2005). *The Effects of Integrating Wind Power on Transmission System Planning, Reliability, and Operations: Report on Phase 2*. Prepared for the New York State Energy Research & Development Authority. Schenectady, New York: GE Energy.

GE Energy. (2008). *Analysis of Wind Generation Impact on ERCOT Ancillary Services Requirements*. Prepared for the Electricity Reliability Council of Texas. Schenectady, New York: GE Energy.

GE Energy. (2010). *Western Wind and Solar Integration Study.* NREL/SR-550-47434. Golden, Colorado: National Renewable Energy Laboratory.

Idaho Public Utilities Commission. (2010). *In the Matter of the Petition of Pacificorp dba Rocky Mountain Power for an Order Revising the Wind Integration Rate for Wind Powered Small Power Generation Qualifying Facilities.* PAC-E-09-07 Order No. 31021. March 12, 2010.

IEEE Power and Energy Magazine. (2009). *An Update on Wind Integration*, 7(6), November-December.

IHS CERA. (2009). *Power Plant Construction Costs Continue to Fall, but Decline is Limited.*

December 18, 2009.

IHS Emerging Energy Research. (2010). *Personal communication with Keith Hayes*. June 4, 2010.

Kirkegaard, J., Hanemann, T. & Weischer, L. (2010). *It Should Be a Breeze: Harnessing the Potential of Open Trade and Investment Flows in the Wind Energy Industry*. WP 09-14. Washington, DC: World Resources Institute.

Martin, K. (2010). *Economics of Wind 102.* Presentation at AWEA's WINDPOWER 2010 Conference & Exhibition, Dallas, Texas, May 24, 2010.

MISO. (2010). *Dispatchable Intermittent Resources.* Market Subcommittee Meeting. January 2010.

Musial, W. & Ram, B. et al. (2010). Large-scale Offshore Wind for the United States: Technical Assessment of Barriers and Opportunities. Forthcoming. Golden, Colorado: National Renewable Energy Laboratory.

North American Electric Reliability Corporation (NERC). (2009a). *Long-Term Reliability Assessment 2009-2018*. October 2009.

North American Electric Reliability Corporation (NERC). (2009b). *Accommodating High Levels of Variable Generation*. North American Electric Reliability Corporation.

Office of the Press Secretary. (2010). *Fact Sheet: $2.3 Billion in New Clean Energy Manufacturing Tax Credits.* January 8, 2010. http://www.whitehouse.gov/the-press-office/fact- sheet-23 -billion-new-clean-energy-manufacturing-tax-credits

PacifiCorp. (2005). *Technical Appendix for the 2004 Integrated Resource Plan*. Portland, Oregon: PacifiCorp.

PacifiCorp. (2007). *Technical Appendix for the 2007 Integrated Resource Plan*. Portland, Oregon: Pacificorp.

PacifiCorp. (2009). *2008 Integrated Resource Plan: Volume II - Appendices*. Portland, Oregon: Pacificorp.

Pfeifenberger, J., Fox-Penner, P. & Hou, D. (2009). *Transmission Investment Needs and Cost Allocation: New Challenges and Models.* Presentation to the Federal Energy Regulatory Commission. December 1, 2009.

Piwko, R., et al. (2007). *Intermittency Analysis Project Appendix B: Impact of Intermittent Generation on Operation of California Power Grid*. Sacramento, California: California Energy Commission, PIER Research Development & Demonstration Program.

Porter, K. & Rogers, J. (2010). *Status of Centralized Wind Power Forecasting in North America*. NREL/SR-550-47853. Golden, CO: National Renewable Energy Laboratory.

Porter, K., Fink, S., Mudd, C. & DeCesaro. J. (2009). *Generation Interconnection Policies and Wind Power: A Discussion of Issues, Problems, and Potential Solutions*. NREL/SR-550-44508. Denver, Colorado: National Renewable Energy Laboratory.

Portland General Electric (PGE). (2009). *Integrated Resource Plan 2009*. Portland, Oregon: Portland General Electric.

Puget Sound Energy. (2007). *2007 Integrated Resource Plan, Appendix G- Wind Integration Studies*. Bellevue, Washington: Puget Sound Energy.

SCE. (2010). *Southern California Edison Celebrates Milestone for a Major Renewable Transmission Project*. May 4, 2010. http://www.edison.com/pressroom/pr.asp?bu=&year=0&id=7390.

Shiu, H., Milligan, M., Kirby, B. & Jackson, K. (2006). *California Renewables Portfolio Standard Renewable Generation Cost Analysis: Multi-Year Analysis Results and*

Recommendations.
Consultant report prepared by the California Wind Energy Collaborative. Sacramento, California: California Energy Commission.

SPP. (2010). *Submission of Tariff Revisions to Modify Transmission Cost Allocation Methodology*. Docket No. ER10-1069-000.

Stolarski, Christopher, Raymond Wood, John Eber, Stephen Schaefer. 2010. *Capital Sources for the Future.* Panel discussion at AWEA's WINDPOWER 2010 Conference & Exhibition, Dallas, Texas, May 24, 2010.

3Tier. (2010). *3TIER Releases Wind Performance Maps: Maps illustrate impact of El Niño in 2009*. February 4, 2010. http://www.3tier.com/en/about/press-releases/3tier-releases-windperformance-maps/

Tonbridge. (2010). *Tonbridge Power Inc. Announces 2010 First Quarter Results.* May 28, 2010. http://www.marketwire.com/press-release/Tonbridge-Power-Inc-Announces-2010-FirstQuarter-Results-1 268293 .htm.

UBS Limited. (2010). *UBS Global I/O: Global Wind Sector.* UBS Investment Research. 15 March.

U.S. Court of Appeals for the Fourth Circuit. (2009). *Piedmont Environmental Council v. FERC,* No. 07-1651 (4th Cir. Feb. 18, 2009).

U.S. Court of Appeals for the Seventh Circuit. (2009). *Illinois Commerce Commission, et al. v. FERC*, Nos. 08-1306, *et al.* (7th Cir. Aug. 6, 2009).

Ventyx. (2010). *Velocity Suite Data Product*. Accessed June 15, 2010.

WAPA. (2009). *Montana-Alberta Tie Limited Project, Western's first TIP Project.* http://www.wapa.gov/recovery

Wyden, R. (2010). *U.S. Trade in Environmental Goods: Follow-Up Report to 'Major Opportunities and Challenges to U.S. Exports of Environmental Goods.'* Washington, DC: United States Senate.

Zaelke, E., Carstensen, C., Markowitz, L., Walsh, K. & Abel, R. (2010). *Optimizing Capital in the Current Environment.* Panel discussion at AWEA's WINDPOWER 2010 Conference & Exhibition, Dallas, Texas, May 24, 2010.

End Notes

[1] The 100 kW cut-off between 'small' and 'large' wind turbines is, in part, justified by the fact that the U.S. tax code makes a similar distinction. AWEA (2010a) sometimes uses this distinction as well, but in other instances defines large wind turbines as including turbines above *and* equal to 100 kW.

[2] For example, large wind turbines are defined in this report as exceeding 100 kW, and by AWEA (2010a) as equal to and exceeding 100 kW. In reporting annual and cumulative capacity additions, this report focuses on large turbines, whereas AWEA (2010a) sometimes also includes small wind turbines. Other methodological differences between AWEA (2010a) and this report are noted as appropriate in the pages that follow.

[3] When reporting annual wind power capacity additions, this report focuses on *gross* capacity additions of large wind turbines. The *net* increase in capacity each year can be somewhat lower, reflecting turbine decommissioning. Some of the methodological differences between the figures presented here and by AWEA (2010a) are summarized in footnote 2. These difference lead AWEA (2010a) to report 10,010 MW of wind power capacity additions in 2009 (including large and small wind turbines), for a cumulative total of 35,086 MW

[4] These investment figures are based on an extrapolation of the average project-level capital costs reported later in this report, and do not include investments in manufacturing facilities, research & development expenditures, or O&M costs.

[5] The same trend is apparent in Europe. In 2009, for example, more wind power was installed in the EU than any other generating technology, with 39% of all capacity additions coming from wind power (EWEA 2010). From 2000 through 2009, 33% of capacity additions in the EU came from wind power, second only to natural gas.

[6] Given the relatively low capacity factor of wind power, one might initially expect that percentage contribution of wind power on an *energy* basis would be much lower than on a *capacity* basis. This is not necessarily the case, as documented by a review of capacity and electricity production data from EIA, in part because even though combined-cycle gas plants can be operated as baseload facilities with high capacity factors, those facilities are often run as intermediate plants with capacity factors that are not dissimilar from that of wind power. Combustion turbine gas facilities run at even lower capacity factors.

[7] Wind power additions in China are from BTM (2010), and include a considerable amount of capacity that was installed but that had not yet received transmission interconnection by the end of 2009. All of the U.S. capacity reported here, on the other hand, was capable of electricity delivery. In fact, if only considering the new wind power capacity that achieved transmission interconnection and was therefore capable of delivering electricity to the grid by the end of 2009, the United States would have again led the world in annual capacity additions in 2009.

[8] Yearly and cumulative installed wind power capacity in the United States are from AWEA, while global wind power capacity in 2009 comes from BTM (2010), but updated with the most recent AWEA data for the United States. Global wind power capacity in earlier years comes from the Earth Policy Institute. Some disagreement exists among these data sources and others, e.g., *Windpower Monthly* and the Global Wind Energy Council.

[9] In terms of actual 2009 deliveries, wind energy represented 1.8% of net electricity generation and 2.0% of national electricity consumption in the United States. These figures are below the 2.5% figure provided above because 2.5% is a projection based on end-of-year 2009 wind power capacity.

[10] "Large-scale" turbines are defined consistently with the rest of this report – i.e., turbines over 100 kW.

[11] Wind energy penetration can either be expressed as a percentage of in-state load or in-state generation. In-state generation is used here, primarily because wind energy is often sold across state lines, which tends to distort penetration levels expressed as a percentage of in-state load. To estimate these figures, end-of-2009 wind power capacity is translated into estimated annual wind electricity production based on estimated state-specific capacity factors that derive from the project performance data reported later in this report. The resulting state-specific wind electricity production estimates are then divided by the latest data on total in-state electricity generation available from the EIA (i.e., 2009). The resulting wind energy penetration estimates shown in Table 2 differ from what AWEA provides in its *U.S. Wind Industry Annual Market Report* (AWEA 201 0a). The most significant source of these differences is that AWEA uses preliminary data on actual 2009 wind electricity generation from EIA, while this report estimates annual wind electricity generation based on the amount of wind power capacity installed at the end of 2009.

[12] In calculating these figures, several issues deserve mention. First, the utility-specific capacity data that AWEA released in its *U.S. Wind Industry Annual Market Report* (AWEA 201 0a) are used, with two exceptions: (1) the Empire District Electric Company, with 255 MW of wind power under contract at the end of 2009, was added to AWEA's "top twenty" investor-owned utility list at position number 14 (and ranks 19th in our combined list of all utility types); and (2) Minnkota Power Cooperative's wind power capacity was corrected to 357 MW (AWEA (2010a) shows 290 MW). Second, only utilities with more than 100 MW of wind power capacity are included in the calculation of wind energy as a proportion of retail sales. Third, projected wind generation based on each utility's installed wind power capacity at the end of 2009 is divided by the aggregate national retail sales of that utility in 2008 (which is the latest full year of utility-specific retail sales data provided by EIA). Fourth, in the case of generation and transmission (G&T) cooperatives and power authorities that provide power to other cooperatives and municipal utilities (but do not directly serve retail load themselves), 2008 retail sales from the electric utilities served by those G&T organizations and power authorities are used. In some cases, these individual utilities may be buying additional wind power directly from other projects, or may be served by other G&T cooperatives or power authorities that supply wind. In these cases, the penetration percentages shown here may be somewhat misleading. As an example, the "MSR Public Power Agency" (MSR) is a joint powers agency created to procure power for municipal utilities in the California cities of Modesto, Santa Clara, and Redding. The 8.4% penetration rate shown in the second column of Table 3 represents MSR's power purchase agreement with the 200 MW Big Horn wind power project in Washington state. Two of the three municipal utilities participating in MSR, however, purchase additional wind energy from other wind power projects. The result is that if one were to look at these three municipal utilities individually rather than as a group through MSR, their penetration rates would be considerably higher than the 8.4% shown in Table 3.

[13] In June 2010, MMS was renamed the Bureau of Ocean Energy Management, Regulation, and Enforcement. For the purpose of this report, we continue to use the name that applied prior to June 2010. "Interim limited leases" authorize data gathering activities, allowing for the construction of meteorological towers on the Outer Continental Shelf to collect site-specific data on wind speed, intensity, and direction.

[14] Also in Massachusetts, the Town of Hull is considering shelving a small, planned offshore project due to cost concerns.

[15] The queues surveyed include PJM Interconnection, Midwest Independent System Operator (MISO), New York ISO, ISO-New England, California ISO, Electricity Reliability Council of Texas (ERCOT), Southwest Power Pool (SPP), Western Area Power Administration (WAPA), Bonneville Power Administration (BPA), Tennessee Valley Authority (TVA), and 23 other individual utilities. To provide a sense of sample size and coverage, the ISOs, RTOs, and utilities whose queues are included here have an aggregated peak demand of almost 70% of the U.S. total. Figures 7 and 8 only include projects that were active in the queue at the end of 2009 but that had not yet been built; suspended projects are not included.

[16] As a rough benchmark, 300 GW of wind power capacity is also the approximate amount of capacity required to reach 20% wind energy penetration in the United States, as estimated in DOE (2008).

[17] Market share reported here is in MW terms, and is based on project installations in the year in question, not turbine shipments or orders.

[18] In addition, U.S. manufacturers are major players in the global market for smaller-scale turbines (AWEA 2010b).

[19] Some of these announcements preceded 2009, or were in early 2010, and so are not included in Figure 10.

[20] Despite the different approaches taken, AWEA (2010a), David (2010), and the analysis presented in this section all focus on wind turbines and components and wind turbine costs. Excluded from all three analyses are foundations, electrical collection and grid interconnection systems, roads, project development costs, and other non- turbine balance-of-plant expenditures. Following the approach used by AWEA (2010a), and unlike David (2010), our analysis emphasizes equipment imports as a fraction of wind turbine equipment-related costs alone.

[21] The Department of Commerce trade data are accessed through the U.S. International Trade Commission's (USITC) DataWeb, which compiles statistics from the Department of Commerce on imports and exports. The statistics can be queried online at: http://dataweb.usitc.gov/. The analysis presented here relies on the 'customs value' of imports as opposed to the 'landed value.' For more information on these data and their application to wind energy, see David (2009, 2010).

[22] Harmonized Tariff Schedule (HTS) 8502.31.0000 – "Wind-powered generating sets." This HTS code includes both utility-scale and small wind turbines. When components are imported *separately* from the nacelle, other tariff provisions apply (see footnote 24).

[23] Using similar HTS codes, data on global imports and exports come from the World Trade Atlas by Global Trade Information Services, Inc.

[24] Estimating separate wind turbine component imports is complicated by the fact that the HTS does not contain codes that are exclusive to wind turbine components. Included in the analysis presented here are: HTS 8501.64.0020 – "AC generators from 750 to 10,000 kVA"; HTS 8412.90.9080 – "parts of engines and motors"; HTS 8503.00.9545 – "parts of generators (other than commutators, stators, and rotors)"; HTS 7308.20.0000 – "towers and lattice masts"; HTS 8483.40.5010 – "fixed ratio speed changers"; and HTS 8543.40.5050 – "speed changers other than fixed ratio."

[25] Specifically, based on a review of the countries of origin for the imports, personal communications with USITC and AWEA staff, David (2010), and Wyden (2010), we assume that 70% of the 750–10,000 kVA AC generators (HTS 8501 .64.0020), 25% of the fixed and multiple/variable ratio speed changers (HTS 8483.40.5010 and HTS 8483.40.5050), 70% of the parts of engines, motors, and generators (HTS 8503.00.9545 and HTS 8412.90.9080), and 95% of the towers and lattice masts (HTS 7308.20.0000) are wind-related components. We assume that these percentages apply equally across the entire 2006-2009 time period.

[26] Over the entire 2006-2009 timeframe, the largest source countries for the trade category that includes towers were: Korea (20%), China (18%), Vietnam (12%), Denmark (11%), and Canada (10%) (in 2009, the top three countries were China (28%), Korea (26%), and Vietnam (11%)). For the trade category that includes blades and other components, the largest source countries from 2006-2009 were: Brazil (15%), Germany (8%), Denmark (7%), Mexico (4%), and India (4%) (in 2009, the top three countries were Brazil (20%), Germany (11%), and Denmark (8%)). Finally, for the trade category that includes generators, the largest source countries from 2006-2009 were Germany (37%), Denmark (19%), Japan (16%), Spain (8%), and China (4%) (in 2009, the top three countries were Germany (32%), Japan (31%), and Denmark (8%)). Not included is a country breakdown for the trade category that includes gearboxes because only 25% of the imports in that category are assumed to represent wind power equipment. Even among the categories highlighted here, source country designations should be viewed with caution since the trade categories are not exclusive to wind energy.

[27] Total wind turbine costs ($/kW) are assumed to equal approximately 75% of the average project-level costs reported later in this report in Figure 27, while wind turbine equipment-related costs are assumed to equal 85% of total wind turbine costs (with the remaining 15% consisting of transportation, project management, and other soft costs). To calculate total calendar-year wind turbine equipment-related costs, we multiply this wind turbine equipment-related cost figure in $/kW by annual wind power capacity installations. Note that David (2010) does not de-rate total wind turbine costs to estimate equipment-related costs alone, and the estimated

import shares reported by David (2010) therefore differ somewhat from those reported here and by AWEA (2010a).

[28] Reporting these figures as a proportion of total wind project installed costs (not just wind turbine equipment-related costs) is also of interest, but is complicated by the fact that non-turbine balance-of-plant costs may also involve some level of imports. Nonetheless, if one simply assumes that 80% of non-turbine-equipment balance-ofplant costs derive from domestic sources with the remaining 20% from imports, then the import fraction for total wind project installed costs would equal 60% in 2006, declining to 53% in 2007, 37% in 2008, and 32% in 2009.

[29] Modest differences exist between these figures and those presented by AWEA (2010a) for the reasons discussed in footnote 2.

[30] Figure 14 (as well as a number of the other figures and tables included in this report) combines data into both one- or two-year periods in order to avoid distortions related to small sample size in the PTC lapse years of 2000, 2002, and 2004; though not a PTC lapse year, 1998 is grouped with 1999 due to the small sample of 1998 projects.

[31] Projects less than 2 MW in size are excluded from Figure 16 so that a large number of single-turbine "projects" (that, in practice, may have been developed as part of a larger, aggregated project) do not end up skewing the average. For projects installed in phases, each phase is considered to be a separate project. Projects that are partially constructed in two different years are counted as coming online in the year in which a clear majority of the capacity was completed. If roughly equal amounts of capacity are built in each year, then the full project is counted as coming online in the later year. Due to methodological differences, these figures differ somewhat from those presented in AWEA (2010a), though the trend is consistent.

[32] Only announced transactions that are believed to involve 500 MW or more of in-development U.S. wind power projects are included. Not included are transactions in which one developer purchases another developer's project pipeline, without an actual acquisition of or investment in that developer.

[33] Hedges are often structured as a "fixed-for-floating" power price swap – a purely financial arrangement whereby the wind power project swaps the "floating" revenue stream that it earns from spot power sales for a "fixed" revenue stream based on an agreed-upon strike price. For some projects (especially where natural gas is virtually always the marginal supply unit), the hedge is structured in the natural gas market rather than the power market, in order to take advantage of the greater liquidity and longer terms available in the forward gas market.

[34] Power marketers are defined here to include not only traditional marketers such as PPM Energy (now part of Iberdrola), but also the wholesale power marketing affiliates of large investor-owned utilities (e.g., PPL Energy Plus or FirstEnergy Solutions), which may buy wind power on behalf of their load-serving affiliates.

[35] Three primary factors significantly restrict the size of this sample: (1) projects located within ERCOT (in Texas) fall outside of FERC's jurisdiction, and are therefore not required to report prices (reduces sample by about 8,600 MW); (2) the increasing number of utility-owned projects are not included, since these projects do not sell their power on the wholesale market (reduces sample by about 5,300 MW); and (3) the increasing number of merchant (or quasi-merchant) projects that sell power and RECs separately are not included in the sample, because the power price reported by these projects only represents a portion of total revenue received (reduces sample by roughly another 4,200 MW). In addition, certain "qualifying facilities" are not required to report their power sales to FERC.

[36] All wind power pricing data presented in this report exclude the few projects located in Hawaii. Those projects are considered outliers in that they are significantly more expensive to build than projects in the continental United States, and receive a power sales price that is significantly higher-than-normal, in part because it has historically been linked to the price of oil.

[37] Prices from two individual projects built during the 2000-200 1 period, and one project built in 2008, are not shown in Figure 20 (due to the scale of the y-axis), but are included in the capacity-weighted averages for those periods. The omitted prices are roughly $95/MWh and $1 50/MWh in the earlier period, and $ 126/MWh for the 2008 project.

[38] Although it may seem counterintuitive, the weighted-average price in 1999 for projects built in 1998 and 1999 (shown in Figure 19 to be about $65/MWh) is significantly higher than the weighted-average price in 2009 for projects built in 1998 and 1999 (shown in Figure 20 to be about $33/MWh) for three reasons: (1) the sample size is larger in Figure 20, due to the fact that 2009 prices are presented, rather than 1999 prices as in Figure 19 (i.e., we were unable to obtain early-year pricing for some of the projects built in 1998-1999); (2) two of the larger projects built in 1998 and 1999 (for which both 1999 and 2009 prices are available, meaning that these projects are represented within both figures) have nominal PPA prices that actually *decline*, rather than remaining flat or escalating, over time; and (3) inflating all prices to constant 2009 dollar terms impacts older (i.e., 1999) prices more than it does more-recent (i.e., 2009) prices.

[39] Average prices in Texas and New England, in particular, may not be representative as those averages include just three and two projects, respectively. Once again, sample size in Texas is severely limited (despite the enormous growth of wind power capacity in that state) because generators located within ERCOT are not required to file pricing information with FERC. As such, the pricing information for Texas provided in this

report comes primarily from projects located in the Texas panhandle, which is within the Southwest Power Pool (SPP) rather than ERCOT. Note also that projects in this area have not experienced the same level of curtailment as is common in ERCOT which, in combination with a strong wind resource in the region and relatively low capital costs (two of the three projects, totaling 75% of the aggregate capacity, were built earlier in the 2006-2009 time period), may have facilitated lower prices than in other parts of Texas. One of the two New England projects in the sample over this period is not shown in Figure 21 because its price ($126/MWh) exceeds the scale of the y-axis; however, this project's price is included in the capacity-weighted average for New England.

[40] A flat block of power is defined as a constant amount of electricity generated and sold over a specified time period. Though wind power projects do not provide a perfectly flat block of power, as a common point of comparison, a flat block is not an unreasonable starting point. In other words, the time-variability of wind energy is often such that its wholesale market value is somewhat lower than, but not too dissimilar from, that of a flat block of (non-firm) power.

[41] The five pricing nodes represented in Figure 22 by an open, rather than closed, bullet do not have complete pricing history back through 2003. As such, the wholesale electricity range presented in Figure 23 does not, in every year, reflect data from the complete set of hubs.

[42] Although its price ($126/MWh) is factored into the capacity-weighted average wind power price (depicted by the red dash), one New England project is not shown in Figure 24, due to the compressed y-axis scale. As discussed in footnote 39, the average wind energy prices for Texas and New England presented here should be viewed with caution.

[43] Operations and maintenance (O&M) costs are another important variable that affects wind power prices. A later section of this report covers trends in project-level O&M costs.

[44] In Figures 25 and 26, three individual project outliers (the same three described earlier in footnote 37) are obscured by the compressed y-axis scale, yet still influence the trend line.

[45] Limited sample size early on – particularly in the 1 980s – makes it difficult to pin down this number with a high degree of confidence.

[46] Learning curves have been used extensively to understand past cost trends and to forecast future cost reductions for a variety of energy technologies, including wind energy. Learning curves start with the premise that increases in the cumulative production or installation of a given technology leads to a reduction in its costs. The principal parameter calculated by learning curve studies is the learning rate: for every doubling of cumulative production/installation, the learning rate specifies the associated percentage reduction in costs. Based on the installed cost data presented in Figure 27 and global cumulative wind power installations, learning rates can be calculated as follows: 9.4% (using data from 1982 through 2009) or 14.4% (using data only during the period of cost reduction, 1982-2004).

[47] It is important to recognize that wind power projects were not alone in seeing upward pressure on project costs – other types of power plants experienced similar increases in capital costs. For example, the IHS CERA Power Capital Cost Index of coal, gas, and wind power plants shows a 74% capital cost increase from 2000 to the end of 2009 (IHS CERA 2009).

[48] For example, data compiled by Berkeley Lab show an estimated weighted-average cost for a sample of more than 4,300 MW of projects likely to be built in 2010 of $2,230/kW, or $1 10/kW higher than for the sample of projects completed in 2009.

[49] Graphical presentation of the data in this way should be viewed with some caution, as numerous factors influence project costs (e.g., whether projects are repowered vs. "greenfield" development, etc.). Actual cost differences among some regions may therefore be more (or less) significant than they appear in Figure 29.

[50] Because of data limitations, the precise content of many of the individual transactions is not known.

[51] Though some performance data for wind power projects installed in 2009 are available, those data do not span an entire year of operations. As such, for the purpose of this section, the focus is on projects with commercial operation dates in 2008 and earlier.

[52] There are fewer individual projects – though more capacity – in the cumulative sample for 2009 than there were in 2008. This is due to the sampling method used by the EIA, which focuses on a subset of larger projects throughout the year, before eventually capturing the entire sample some months after the year has ended. As a result, it might be late 2010 before the EIA reports 2009 performance data for all of the wind power projects that it tracks, and in the mean time, this report is left with a smaller sample consisting mostly of the larger projects in each state.

[53] Data on curtailment in other regions were not available, but are expected to be relatively modest for 2009.

[54] Although focusing just on 2009 tends to limit the effects of inter-annual fluctuations in the nationwide wind resource (which do impact the year to year results in Figure 31), it also means that the *absolute* capacity factors shown in Figure 33 may not be representative if 2009 was not a representative year in terms of the strength of the wind resource. Note also that by including only 2009 capacity factors, variations in the quality of the wind resource year in 2009 across regions could skew the regional results presented in Figure 34 and Table 8.

[55] Projects installed in 2009 are not shown because only data from the first full year of project operations (and afterwards) are used, which in the case of projects installed in 2009 would be year 2010 (for which data are not yet available).

[56] Many of the projects installed more-recently may still be within their turbine manufacturer warranty period, in which case the O&M costs reported here may or may not include the costs of the turbine warranty, depending on whether the warranty is paid up-front as part of the turbine purchase, or is paid over time.

[57] Figure 36 shows that, among projects in their third year of commercial operation, the most recent vintage of projects (i.e., those installed between 2006 and 2008) had the *highest* average O&M costs. This is the result, however, of a single project completed in 2006 with particularly high O&M costs in its third year of operation.

[58] Such entities have also been eligible to receive the Renewable Energy Production Incentive, which nominally offers a 10-year cash payment equal in face value to the PTC, but the need for annual appropriations and insufficient funding under those appropriations has limited the availability of these payments and therefore also the effectiveness of the program.

[59] The $800 million portion of the $2.4 billion authorization that was reserved for rural electric cooperatives was undersubscribed by $200 million, which is why only $2.2 billion was allocated.

[60] For example, Xcel Energy submitted a resource zone and transmission plan to the Colorado Public Utilities Commission outlining five wind and solar resource zones, three transmission projects currently being implemented, and six potential transmission projects that would greatly increase renewable energy transfer capability from the zones. The Western Renewable Energy Zones initiative, a collaborative between the Western Governors' Association and the U.S. Department of Energy, completed its Phase 1 report in June 2009. During Phase 1, the project developed a methodology to identify and characterize specific renewable resource-rich areas and created a modeling tool to evaluate the relative economic costs of delivering the renewable energy from the zones to load centers. Arizona's three largest utilities, meanwhile, filed proposals with the Arizona Corporation Commission to build three renewable-oriented transmission projects each. A variety of other state-level renewable energy zone analyses also continued in 2009.

[61] Wind penetration on a capacity basis (defined as nameplate wind power capacity serving a region divided by that region's peak electricity demand) was frequently used in earlier integration studies. For a given amount of wind power capacity, penetration on a capacity basis is typically higher than the comparable wind penetration in energy terms (because, over the course of a year, wind power projects generally operate at a lower percentage of their rated capacity, on average, than do many other resources). The energy penetration levels in the EWITS study correspond to 48% wind on a capacity basis. The energy penetration levels in the WWSIS study correspond to penetrations of 52% wind and 10% solar in the WestConnect and 38% wind and 6% solar in the rest of the Western Electricity Coordinating Council on a capacity basis.

[62] The integration costs considered in these studies typically refer to the costs associated with accommodating the variability and uncertainty associated with wind energy. Generally, these costs are associated with three different time frames: *regulation* – from seconds to a few minutes; *load-following* – tens of minutes to a few hours; and *unit commitment* – out to the next day or two. Studies often, but not always, estimate these costs as the difference in overall electric system production costs between a scenario that captures the variability and unpredictability of wind energy and a scenario with an energy-equivalent block of power having no variability or uncertainty.

[63] Several additional studies focus on the operational impacts of wind energy and/or on the overall production-cost reduction value of wind energy, without an explicit comparison to an energy-equivalent block of power having no variability or uncertainty. These studies are not included in the table because they do not seek to explicitly calculate integration costs. Examples of such studies include: the Western Wind and Solar Integration Study, (GE 2010); a 2010 study of wind energy integration into the Southwest Power Pool (Charles River Associates 2010); a 2008 study on the ancillary service implications of high wind energy penetration in ERCOT (GE 2008); two integration studies for California from 2007 by the California ISO (CAISO 2007) and by the California Energy Commission's Intermittency Analysis Project (Piwko et al. 2007); and a study for New York in 2005 (GE 2005).

[64] Some of the studies included in the table also address capacity valuation for resource adequacy purposes; those results are not presented here.

[65] The relatively low cost estimates in the 2006 Minnesota study and the 2010 Nebraska study, despite aggressive levels of wind energy penetration, are partly a result of relying on the broader regional electricity market to accommodate certain elements of integrating wind energy into system operations. Conversely, the higher integration costs found by Avista and Idaho Power are, in part, caused by the relatively smaller markets in which the wind energy is being absorbed and by those utilities' operating practices. Specifically, the Northwest currently uses hourly scheduling intervals rather than the sub-hourly markets common in ISOs and RTOs. A sensitivity case in the Avista Utilities study demonstrates that the use of a 10-minute transaction scheduling interval would decrease the cost of integrating wind energy by 40-60%.

[66] Two estimates of integration costs from 2009 are not included in this table due to ongoing efforts to refine the methodologies and/or incomplete documentation those methodologies. The first study, from PacifiCorp, estimated integration cost to be $11 .85/MWh at a 22% wind power capacity penetration level, assuming a

$45/ton CO2 tax. With a lower CO2 tax of $8/ton, estimated integration costs decreases to $9.96/MWh (PacifiCorp 2009). The second study, from Portland General Electric, estimated integration costs to be $11 .75/MWh at a 27% wind power capacity penetration level (PGE 2009). If and when refined estimates and more complete documentation of study methodologies become available, these results will be included in the main body of this report..

In: U.S. Wind Power Industry Market Conditions and Trends ISBN: 978-1-61209-559-2
Editor: Liam E. Dempsey

Chapter 2

IMPACT OF WIND ENERGY INSTALLATIONS ON DOMESTIC MANUFACTURING AND TRADE

Andrew S. David*

ABSTRACT

This paper analyzes the impact of the increase in wind turbine demand from 2005 to 2009 on U.S. production and trade. The results of this analysis indicate that wind turbine imports rose substantially, but not as fast as U.S. production. Rising demand for wind turbines led to both a significant expansion of domestic production capacity and an increase in imports through 2008. The financial crisis and recession, however, depressed both domestic production and wind turbine imports in 2009. Overall, imports peaked as a share of the market in 2006 and U.S. production in 2008 and 2009 was significantly higher than in 2005, indicating a growing role for domestic producers. The number of foreign and domestic wind turbine original equipment manufacturers (OEMs) producing nacelles or blades at one or more locations in the United States increased from one in 2005 to nine by the end of 2009, with six more OEMs planning to begin U.S. production. Other companies increased investment in U.S. wind turbine blade and tower production. If planned manufacturing plants come online in the next few years, U.S. capacity will continue to expand.

INTRODUCTION

The Obama administration has prioritized expansion of renewable energy as a way to reduce greenhouse gas emissions and revitalize the economy. Central to this strategy is investment in wind energy, which is already adding thousands of megawatts (MW) each year

* This paper represents solely the views of the author and is not meant to represent the views of the U.S. International Trade Commission or any of its commissioners. The invaluable assistance of Michael Anderson, Dennis Fravel, Peg Hausman, Deborah McNay, and Monica Reed is gratefully acknowledged. Please direct all correspondence to Andrew David, Office of Industries, U.S. International Trade Commission, 500 E Street, SW, Washington, DC 20436, telephone: 202-205-3368, fax: 202-205-2018, email: andrew.david@usitc.gov.

to the U.S. electric grid. Yet there is debate about the economic impact of wind energy investment and the extent to which the U.S. industry will benefit from the growth in wind turbine installations. This paper thus analyzes the impact of the increase in wind turbine demand from 2005 to 2009 on U.S. production and trade.

The results of this analysis indicate that U.S. wind turbine imports rose substantially, but not as fast as domestic production.[1] Rising demand for wind turbines led to both a significant expansion of domestic production capacity and an increase in imports through 2008. The financial crisis and recession, however, depressed both domestic production and wind turbine imports in 2009. Overall, imports peaked as a share of the market in 2006 and U.S. production in 2008 and 2009 was significantly higher than in 2005, indicating a growing role for domestic producers. If planned U.S. manufacturing plants come online in the next few years, U.S. production capacity will continue to expand.

This paper focuses on the manufacturing and trade of utility-scale (>100 kilowatts) wind turbines from 2005 to 2009.[2] It analyzes import and export trends, the extent to which U.S. production is growing, U.S. production as a share of the wind turbine equipment market in the United States, and employment and wages in the industry. The primary focus of this paper is on the main components of a wind turbine: the nacelle, blades, and tower (box 1). There will also be a short discussion of the value of gearbox and generator imports.

BOX 1. WIND TURBINE COMPONENTS

Blade: Most wind turbines have three blades. Blades are generally 30 to 50 meters (100 to 165 feet) long, with the most common sizes around 40 meters (130 feet).

Gearbox: Many turbines have a gearbox that increases the rotational speed of the shaft. Some turbines use direct drive generators that do not require a gearbox.

Generator: Wind turbines typically have an AC generator (housed in the nacelle) that converts the mechanical energy from the wind turbine's rotation into electrical energy.

Nacelle: The nacelle houses the main components of the wind turbine, such as the controller, gearbox, generator, and shafts.

Tower: Towers are usually tubular steel structures around 80 meters (260 feet) high. They consist of several sections of varying heights.

Photo courtesy of U.S. Department of Energy (DOE)/National Renewable Energy Lab (NREL).
Photographer: Lee Fingersh.
Sources: This text box from U.S. International Trade Commission (USITC), *Industry and Trade Summary: Wind Turbines*, June 2009, 2.

BACKGROUND

Global wind turbine installations are rapidly rising, due to the maturation of wind technology and the proliferation of government policies that support the development of wind energy.[3] Global wind turbine installations increased from 11,531 megawatts (MW) per year in 2005 to 37,466 MW per year in 2009, and cumulative installed capacity reached 157,899 MW at the end of 2009.[4] The wind turbine market is diversifying geographically as well. In the 1990s and early 2000s, it generally centered on Europe, but in 2009 installations in Asia and North America exceeded installations in Europe. In 2009, installations in Asia were 14,639 MW; in North America, 10,872 MW; and in Europe, 10,526 MW.[5]

The traditionally large size of the European market is one of the key reasons for the strong position of Europe-based original equipment manufacturers[6] (OEMs) in the wind turbine manufacturing industry.[7] In 2009, at least two Europe-based OEMs were among the top three suppliers of wind turbines in eight of the ten largest markets. Vestas (based in Denmark) and Enercon (Germany) supplied wind turbines to the largest number of markets. U.S.-based General Electric Co. (GE) was the world's second largest supplier in 2009, primarily due to its strong presence in the U.S. market (an estimated 84 percent of turbines supplied by GE were to the U.S. market). However, GE and many European OEMs lost global market share in 2009 as a result of the rapid growth of the Chinese market and the dominant positioning of China-based manufacturers in that market. Two of the five leading wind turbine suppliers in 2009 were China-based OEMs and five of the fifteen leading suppliers were based in China. These OEMs have had limited sales outside of China and their increase in market share to this point is almost exclusively the result of the growth in the domestic market, which accounted for an estimated 99.8% of turbines supplied by the five largest Chinese OEMs.[8] This may change in the next few years since many Chinese OEMs are actively seeking to enter foreign markets.[9]

U.S. wind turbine installations increased every year from 2005 to 2009, and the United States was the largest wind turbine market in the world in terms of annual wind turbine installations from 2005 to 2008 (figure 1). (In 2009, China surpassed the U.S. in annual installations.) Annual U.S. wind turbine installations increased from 2,420 MW in 2005 to 9,922 MW in 2009, and total cumulative wind installations surpassed 35,000 MW by the end of 2009.[10] However, record 2009 installations mask significant instability in the market. The financial crisis and recession contributed to a slowdown in project development starting in the fall of 2008. As a result, there were high installations in the first quarter of 2009 as projects started in the previous year reached completion and high installations in the fourth quarter after provisions in the American Recovery and Reinvestment Act of 2009 (Stimulus Bill) were finalized.[11] There continue to be challenges in the market in 2010 and U.S. installations are projected to decrease: IHS Emerging Energy Research forecasts 2010 installations of 6,300 to 7,100 MW and BTM Consult forecasts installations of 8,000 MW.[12]

Expanding Nacelle, Blade, and Tower Manufacturing in the United States

The growth in wind turbine installations and a period of stability in government policy led to significant investment in U.S. manufacturing by both U.S. and foreign companies from

2005 to 2008, along with a rapid increase in domestic production. In 2009, however, there were fewer announced investments in new manufacturing facilities. Furthermore, while production in 2009 across the key segments of the value chain discussed here (nacelles, blades, and towers) was still significantly higher in 2009 than in 2005, some U.S. producers reduced their workforces and production levels to be in line with reduced market demand due to the financial crisis, a decline in wind turbine orders, and the cancellation or deferral of existing orders.

Rising U.S. Production by OEMs

OEM production capacity in the United States rapidly increased from 2005 to 2009 as companies opened new manufacturing facilities. In 2005 GE was the only OEM with production in the United States, but by 2009 there were nine OEMs producing nacelles, blades, and/or towers in the United States, with six more OEMs planning to open their first U.S. plant. However, as noted earlier, starting in the fall of 2008 OEMs in the U.S. market were significantly affected by the financial crisis and the recession. The downturn's impact varied by manufacturer, but in general it affected OEMs in the following ways:

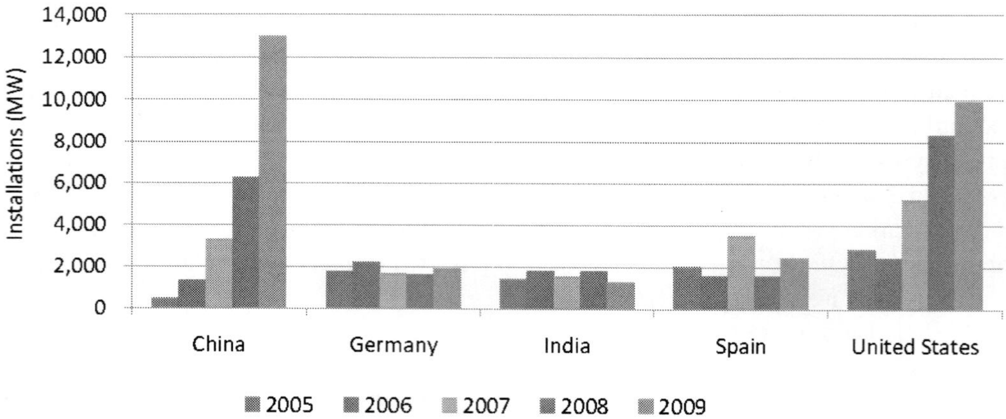

Source: Global Wind Energy Council (GWEC), *Global Wind 2008 Report*, April 2009; GWEC, "Global Wind Power Boom Continues," February 3, 2010.

Figure 1. Annual wind turbine installations 2005–09, top five countries in 2009 installations.

- Starting in the fall of 2008, most OEMs suffered from a lack of new U.S. orders due to the inability of project developers to find financing for wind projects.[13]
- Some (but not all) OEMs had significant wind turbine orders cancelled or deferred.[14]
- The timing of 2009 installations was unusual, with many installations in the first quarter, as projects far along in the construction process moved forward, and many in the fourth quarter as well, after rules for the Stimulus Bill's provisions were finalized. Installations in the second and third quarters were much fewer.[15]
- Some companies increased production in expectation of higher and/or more stable demand in 2009 and were left with excess production capacity and inventory, including turbines that were produced for cancelled or deferred orders and had to be put back onto the market.[16] On the other hand, some new entrants found it easier to

build their supply chain for components due to less tightness in the overall supply chain.[17]

- Some companies had more installations in 2009 than in 2008, but had less production or fewer shipments in 2009, as some of the turbines installed in 2009 were likely shipped in 2008. For example, U.S. installations of GE wind turbines increased from 2,438 wind turbines (3,657 MW) in 2008 to 2,663 wind turbines (3,995 MW) in 2009 (a 9 percent growth rate).[18] However, GE's global deliveries of wind turbines, which are likely closely linked to trends in the U.S. market, decreased from 3,240 turbines in 2008 to 2,633 turbines in 2009 (a 19 percent drop).[19] U.S. installations of Clipper wind turbines increased from 2008 to 2009, but U.S. production decreased from 722 MW to 260 MW.[20] Vestas wind turbine installations in the United States increased from 1,120 MW in 2008 to 1,488 MW in 2009, but wind turbine deliveries to the United States decreased from 1,345 MW to 749 MW during the same two years.[21] In total, Emerging Energy Research reported that 1.5 gigawatts (GW) of wind turbines were delivered in 2008 but installed in 2009.[22]

These market conditions led some OEMs to reduce production levels to be in line with orders, layoff workers, and/or slow the construction of new manufacturing plants.[23] The downturn also affected suppliers of key components such as blades and towers, as will be discussed below. However, far more OEMs were producing in 2009 than in 2005, production was significantly higher than in 2005, most planned manufacturing plants moved forward, and several more OEMs announced or began significant planning for new manufacturing plants in 2009 and early 2010.[24] The United States is likely to continue to be one of the largest global markets in the next few years and many OEMs want to be well positioned in this market. As a result, challenging market conditions in 2009 do not appear to have halted the trend toward increased U.S. production by OEMs.

U.S.-based OEMs, which accounted for 46 percent of U.S. wind turbine installations in 2009, have a significant domestic presence in nacelle production.[25] U.S.-based OEMs generally do not produce blades or towers in-house. GE has three U.S. plants that assemble nacelles and is the largest U.S. producer of nacelles (table 1).[26] U.S.-based Clipper Windpower began production of nacelles in 2006 and was the 13th largest global wind turbine supplier in 2009.[27] Northern Power Systems, a leading U.S.-based producer of small wind turbines, developed a utility-scale turbine that it plans to launch in 2010.[28]

Table 1. OEM production capabilities, current and planned, June 2010

OEM	2009 MW Installed	U.S. Production			
		Nacelles	*Blades*	*Towers*	*Components*
U.S.-based OEMs					
Clipper Windpower[a]	605	X			
General Electric Co. (GE)	3,995	3X			2X
Northern Power Systems[b]	–	P			
Non-U.S.-based OEMs (headquarters)					
A-Power (China)[c]	–	P			
Acciona Windpower (Spain)	204	X			
Alstom (France)	–	P			
DeWind (Korea)[d]	6	X			
Fuhrländer (Germany)[e]	3	P			
Gamesa (Spain)	600	X	X		
Mitsubishi Heavy Industries (Japan)	751	P			
Nordex (Germany)	63	P	P		
Nordic Windpower (United Kingdom)	–	X			
Siemens (Germany)[f]	1,162	P	X		X
Suzlon (India)	702		X		
Vestas Wind Systems (Denmark)	1,488	X	X/P	X	
X = Existing manufacturing facility					
P = Planned manufacturing facility					
3 = Number of manufacturing plants (if more than one)					

Source: Manufacturing plant information compiled from press releases, media reports, company documents and Web sites, AWEA publications, and White House Web site (for the American Reinvestment and Recovery Act of 2009 [ARRA], Advanced Energy Manufacturing Tax Credit [MTC] list of awardees). Installation data from Belyeu, "Status of the U.S. Wind Power Market," May 25, 2010.

Notes: Only includes OEMs with prior installations of small or utility scale wind turbines. There are other U.S.-based companies (e.g., Continental Wind Power) that have announced plans to begin commercial production of utility-scale turbines for the first time. Component plants include only facilities that are separate from blade, tower, and nacelle plants. MW installed includes only utility-scale installations in the United States. Emergya Wind Technologies planned to produce nacelles and blades in Arkansas, but these plans were put on hold in late 2009.

[a] In December 2009, United Technologies Corporation agreed to purchase a 49.5 percent stake in Clipper.

[b] Nacelles for Northern Power Systems turbines will be produced by Merrill Technologies Group in Michigan.

[c] A-Power Energy Generation Systems, U.S. Renewable Energy Group, and American Nevada Co. announced plans in March 2010 to build a wind turbine manufacturing plant in southern Nevada, though a specific site has not yet been selected. While limited details are available, previous statements indicated that A-Power intended specifically to build a nacelle manufacturing plant in the United States.

[d] In August 2009, U.S.-based Composite Technology Corporation (CTC) announced that it reached an agreement to sell DeWind to Daewoo Shipbuilding & Marine Engineering Company (DSME). DeWind nacelle production is contracted to TECO Westinghouse Motor Company in Texas.

[e] While the status of Fuhrländer's plant is uncertain, Fuhrländer was an American Reinvestment and Recovery Act (ARRA) Advanced Energy Manufacturing Tax Credit (MTC) awardee and planning for the plant appears to be moving forward.

[f] Siemens subsidiary Winergy produces components in the United States.

Investment in the U.S. market by foreign-based OEMs significantly increased during this time period, and there are currently seven foreign OEMs with a combined ten operational plants producing blades, nacelles, and towers. Five additional foreign-based OEMs are planning to manufacture in the United States and a total of eight more plants are planned by foreign-based OEMs.[29] OEMs are localizing production in the United States in order to take advantage of the growing market, reduce transportation costs, minimize the risks associated with currency fluctuations, ease logistical challenges associated with exporting large nacelles and components, and avoid import duties.[30] Up to this point foreign-based OEMs have opened a similar number of nacelle and blade plants, with only one OEM currently operating a tower plant. However, subsequent investments are more heavily weighted toward nacelle plants, with seven nacelle plants and only two blade plants planned. The companies investing in U.S. production include both companies with limited U.S. market presence and companies with a large presence in the U.S. market, like Siemens and Vestas (which accounted for a combined 27 percent of the U.S. market in 2009).[31]

Expanding Blade and Tower Manufacturing by Suppliers to OEMs

Blade suppliers made significant investments in wind turbine blade manufacturing from 2005 to 2008, but were negatively impacted by the downturn in wind turbine demand starting in the fall of 2008.[32] Currently, three U.S.-based suppliers (Knight and Carver, Molded Fiberglass, and TPI Composites) and Denmark-based supplier LM Glasfiber, which is the largest independent blade supplier globally, produce blades at a combined total of seven locations in the United States (table 2). Energy Composites announced plans in 2009 to open a manufacturing plant in Wisconsin, and Michigan manufacturer Energetx Composites reached a licensing agreement with Spanish company Aeroblade and was an American Reinvestment and Recovery Act (ARRA) Advanced Energy Manufacturing Tax Credit (MTC) awardee in January 2010 (box 2).[33] These plants are in addition to the four current OEM blade manufacturing plants and two facilities planned by OEMs.

Table 2. U.S. blade production facilities, current and planned, excluding OEMs, February 2010

Company	U.S. Production
U.S.-based Manufacturers	
Energetx Composites	P
Energy Composites Corp	P
Knight and Carver	2X
Molded Fiberglass	2X
TPI Composites	X
Non-U.S.-based OEMs (headquarters)	
LM Glasfiber (Denmark)	2X
X = Existing manufacturing facility	
P = Planned manufacturing facility	
2 = Number of manufacturing plants (if more than one)	

Source: Compiled from press releases, media reports, company documents and Web sites, AWEA publications, and the White House Web site (for the ARRA MTC list of awardees).

Notes: TPI Composites received an ARRA MTC award to build a new blade manufacturing plant, though the plant is not listed here because TPI has not yet announced a production location. Energetx signed a wind turbine blade licensing agreement with Spanish company Aeroblade and was announced as an MTC awardee in January 2010.

BOX 2. ADVANCED ENERGY MANUFACTURING TAX CREDIT (MTC)

The American Reinvestment and Recovery Act (ARRA) or Stimulus Bill authorized up to $2.3 billion for an Advanced Energy Manufacturing Tax Credit (MTC). This is a 30 percent tax credit for investments in new or existing facilities for the production of advanced energy technologies, such as renewable energy, energy storage, energy transmission, energy conservation, and electric vehicles. MTC awardees were announced in January 2010.

Sources: DOE Web site, http://www.energy White House, "Fact Sheet: $2.3 Billion in New Clean Energy Manufacturing Tax Credits," January 8, 2010.

The production of blades in the United States was negatively affected by the financial crisis and recession. The downturn also likely contributed to a slowing in the pace of announced new investments in manufacturing plants. Including OEMs, only one new U.S. manufacturing plant was announced in 2009, compared to five new manufacturing plants announced in 2007 and three in 2008.[34] At least six of the eight companies producing blades in the United States, including OEMs, either laid off workers, furloughed workers, or temporarily halted production at some point after the onset of the financial crisis, though some suppliers subsequently returned to prior staffing levels and some applied for the ARRA MTC in order to help increase production levels.[35]

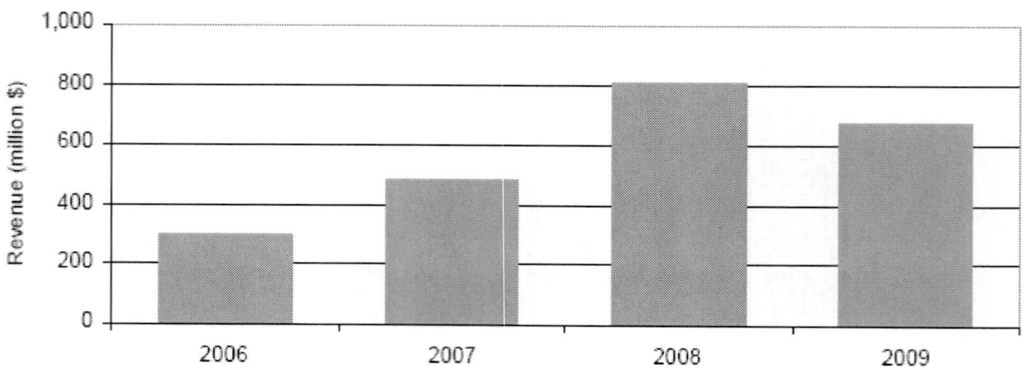

Source: Compiled from company statements and annual reports.
Notes: See footnote 42 in text.

Figure 2. Estimated revenue from tower production of four U.S.-based publicly traded tower manufacturers, 2006–09.

Tower manufacturing in the United States also expanded significantly during this time period, with companies announcing new manufacturing plants, entering the industry for the first time, and increasing revenue and production.[36] The number of tower plants in the United States increased from six in 2004 to 20 in 2009.[37] A number of additional investments in tower manufacturing facilities are planned, with over $70 million of the ARRA MTC funds awarded to companies to manufacture towers.[38] Most tower manufacturers in the United States are U.S.-based companies. Only one OEM, Vestas, currently has a tower manufacturing facility in the United States. This is not surprising, given that OEMs tend to produce a lower percentage of wind turbine towers in-house than blades.[39]

The same trends that affected wind turbine blade producers affected tower producers, with some companies facing reduced orders, cutting production levels, and laying off workers due to the financial crisis and recession.[40] Revenue from sales of towers by four U.S.-based producers—with nine American, one Canadian, and one Mexican production plant among the four—increased from an estimated $302 million in 2006 to $809 million in 2008, then decreased to $679 million in 2009 (figure 2). Yet despite this 16 percent drop in revenue from 2008 to 2009, 2009 revenues were still 125 percent higher than in 2006. In addition, revenue did not decrease for all companies. Revenue declined at two companies, stayed roughly even at the third company, and increased at the fourth company.[41]

Increasing U.S. and Foreign Investment

Annual announced foreign investment in new U.S. nacelle, blade, and tower manufacturing facilities rose from about $39 million in 2005 to $700 million in 2008 (an increase of almost 1,700 percent), then decreased to about $170 million in 2009 due, at least in part, to the financial crisis and recession (figure 3). Total announced foreign investment in new manufacturing plants over these five years was about $1.3 billion.[42] Investment by domestic companies was also significant but is more difficult to quantify, since many of these investments may involve retooling or expanding existing plants rather than opening new ones. U.S. companies manufacture nacelles at four plants and blades at five plants, with at least one more nacelle manufacturing location and two more blade plants planned. In addition, U.S. companies account for most tower manufacturing in the United States.

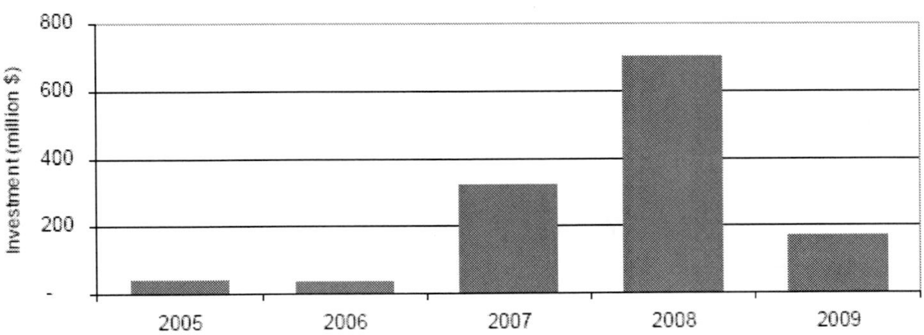

Source: Compiled from press releases, media reports, FDIMarkets investment database (accessed November 3, 2008), and Bloomberg New Energy Finance database (accessed March 12, 2010).
Note: See footnote 43 in text.

Figure 3. Annual announced foreign investment in new U.S. nacelle, blade, and tower manufacturing plants, 2005–08.

U.S. Import Trends

U.S. wind turbine imports substantially increased after 2005, peaking in 2008 and declining slightly in 2009. U.S. imports of wind-powered generating sets—the provision in the Harmonized Tariff Schedule of the United States (HTS) that includes nacelles and, if they are imported with the nacelle, other components (box 3)—increased from $482.5 million in 2005 to $2.5 billion in 2008, then decreased to $2.3 billion in 2009 (figure 4).[43] Denmark was the leading source of imports in 2009, accounting for 34.5 percent of imports. Japan accounted for 25.5 percent, Spain 13.3 percent, India 10.7 percent, and Germany 8.9 percent.[44] Each of these countries is home to a significant wind manufacturing industry.

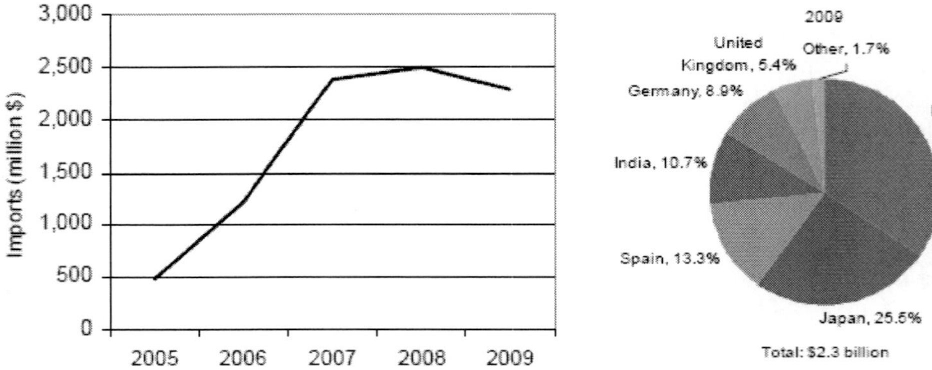

Source: Compiled from official statistics of the U.S. Department of Commerce.

Figure 4. U.S. imports of wind-powered generating sets (HTS 8502.31.0000).

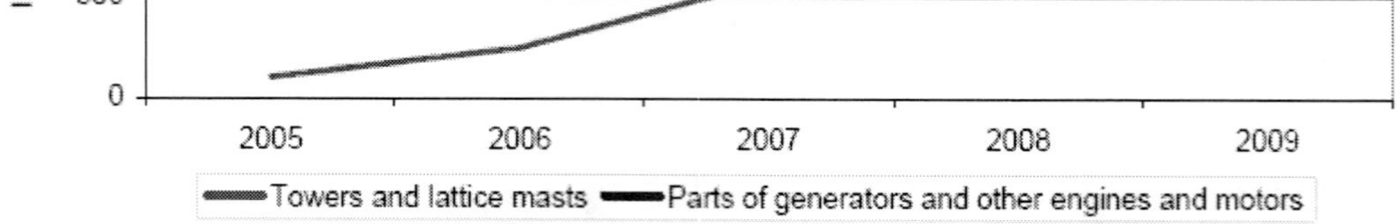

Source: Compiled from official statistics of the U.S. Department of Commerce.

Figure 5. U.S. imports of parts of generators and other engines and motors (HTS 8503.00.9545 and 8412.90.9080) and towers and lattice masts (HTS 7308.20.0000), 2005–09.

BOX 3. HTS NOMENCLATURE FOR WIND TURBINES

Wind turbine nacelles are currently included in 8502.31.0000, wind-powered generating sets, in the Harmonized Tariff Schedule of the United States (HTS). This provision also includes components, such as blades, hubs, and towers, when imported with the nacelle. This HTS provision includes both small and utility-scale wind turbines, but small wind turbines generally represent a low percentage of wind turbine trade. When imported separately, blades and other components, such as hubs, are included in HTS 8412.90.9080, parts of other engines and motors, and HTS 8503.00.9545, parts of generators. Towers are included in HTS 7308.20.0000, towers and lattice masts.

The term "wind-powered generating sets" will be used to refer to trade specifically in HTS 8502.31.0000, while "wind turbine" will be used to refer to installations, production, and trade in more generic terms or in reference to combined trade of generating sets, blades, and towers.

Similarly, U.S. imports in the HTS provisions that include towers and blades increased from 2005 to 2008, then declined in 2009 (figure 5). [45] Imports of other parts of generators and other engines and motors, which includes wind turbine blades and other components such as hubs, increased from $562.7 million in 2005 to $1.8 billion in 2008, then declined to $1.3 billion in 2009. The top five sources of these imports in 2009 were Brazil (21.2 percent of imports), Mexico (17.5 percent), Germany (14.1 percent), India (8.6 percent), and Denmark (8.1 percent).[46]

Imports of towers and lattice masts, the HTS provision that includes wind turbine towers, increased from $102.7 million in 2005 to $944.4 million in 2008, then declined to $612.2 million in 2009.[47] Since 2007 there has been a shift in the sourcing of tower imports from Europe to Asia, with the share of imports from five Asian countries (China, India, Indonesia, Korea, and Vietnam) increasing from 31 percent to 81 percent of imports (figure 6). Imports from Canada and Mexico increased from 2006 to 2008, but declined in 2009 when imports from Canada fell from $152.4 million in 2008 to $15.7 million in 2009 and imports from Mexico from $62.2 million to 43.0 million.[48]

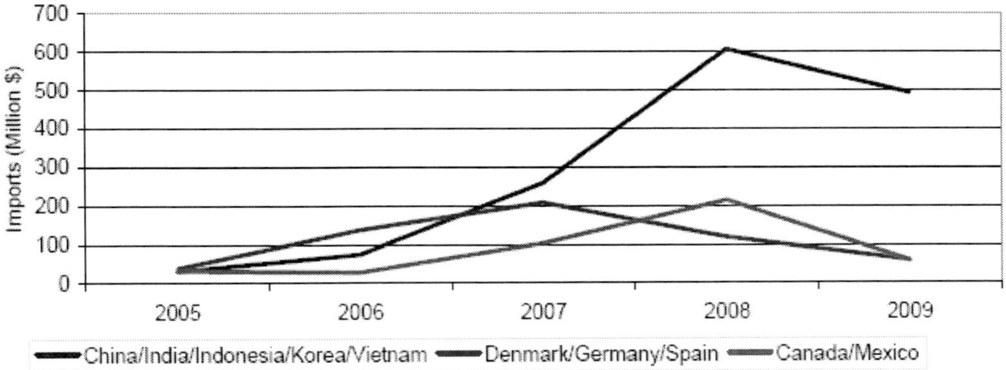

Source: Compiled from official statistics of the U.S. Department of Commerce.

Figure 6. U.S. imports of towers and lattice masts (HTS 7308.20.0000), 2005–09, top 10 countries.

U.S. Production Growth Begins to Outpace Import Growth

The prior two sections indicate both rising production and rising imports of wind turbines, though both were depressed in 2009 by the financial crisis and the recession. The following section assesses the balance between imports and domestic production in the market by comparing imports to (1) the number of MW supplied to the U.S. market and (2) the value of the U.S. market. This analysis indicates that imports peaked as a share of the market in 2006 and subsequently declined to around 2005 levels. Domestic production significantly increased after 2006, capturing a greater share of the growing U.S. market (box 4).

Box 4. Domestic Production

For the purposes of this paper, the terms U.S. production and domestic production include production at all plants in the United States regardless of whether they are owned by U.S.-based or foreign-based companies.

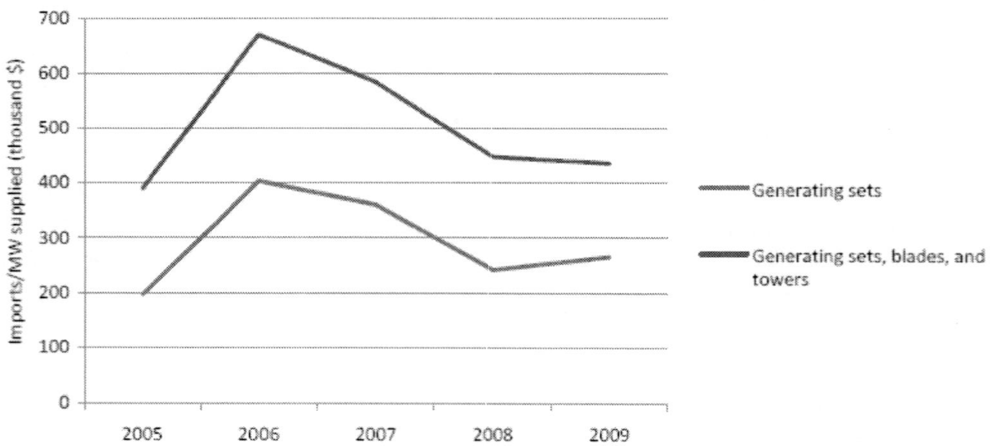

Sources: USITC calculations based on import data compiled from official statistics of the U.S. Department of Commerce and data on wind turbines supplied from BTM Consult.

Note: This figure does not control for changes in wind turbine prices from 2005 to 2009. Import data are based on the full value of imports of wind-powered generating sets (HTS 8502.31.0000) and towers and lattice masts (HTS 7308.20.0000) and two-thirds of the value of imports of parts of generators and other engines and motors (HTS 8503.00.9545 and 8412.90.9080).

Figure 7. Imports of wind-powered generating sets, alone and in combination with estimated imports of towers and blades, per MW of wind turbines supplied, 2005–09.

Imports per MW Supplied to the U.S. Market

U.S. imports of wind-powered generating sets and combined imports of wind-powered generating sets, blades, and towers per MW of wind turbines supplied to the U.S. market peaked in 2006 and subsequently declined significantly.[49] The estimated import value of wind-powered generating sets per MW supplied rose from $196,000 in 2005 to $403,000 in 2006, then declined to $265,000 in 2009 (figure 7).[50] Similarly, the value per MW supplied of combined imports of wind-powered generating sets, blades, and towers increased from an estimated $390,000 per MW in 2005 to $670,000 in 2006 before declining to $435,000 in 2009. Since the market expanded substantially from 2005 to 2009 and the value of imports per MW supplied (of generating sets, blades, and towers) was only slightly higher than in 2005, this indicates a significant increase in domestic production.

Imports as a Share of the Value of the Market

Accounting for the increase in the value of wind turbines over time by calculating imports as a share of the U.S. market (box 5), combined imports of wind-powered generating sets, blades, and towers increased from 38 percent of the market in 2005 to 64 percent of the market in 2006, but declined to 34 percent of the market in 2008 and 32 percent in 2009 (figure 8).[51] These numbers are based on estimates of the value of the wind turbine market and may not be an exact percentage of domestic content, but they do serve as an indicator of the trend in imports as a share of the market. In addition, these data indicate that while domestic production as a share of the market was only slightly higher in 2005 than in 2009, the value of domestic production was significantly higher, since the value of the market increased more than 350 percent.

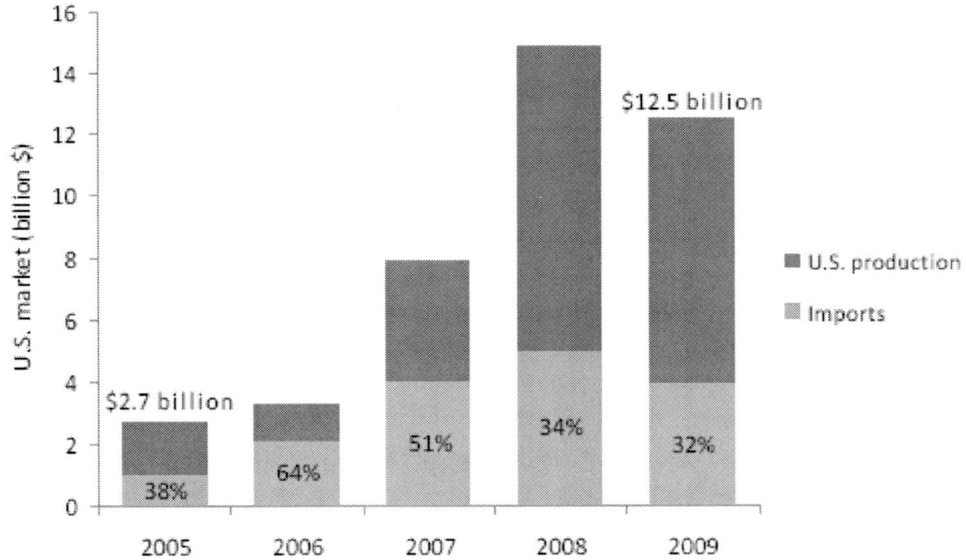

Sources: Import data compiled from official statistics of the U.S. Department of Commerce; value of the U.S. market based on USITC calculations using data on MW supplied from BTM Consult.

Notes: Value of the market calculated based on wind turbines supplied. Total value of wind turbines supplied includes both imports and U.S. wind turbine production. The figure does not include exports. Import data are based on the full value of imports of wind-powered generating sets (HTS 8502.31.0000) and towers and lattice masts (HTS 7308.20.0000) and two-thirds of the value of imports of parts of generators and other engines and motors (HTS 8503.00.9545 and 8412.90.9080). Import data are the landed duty-paid values. See box 5 for a fuller discussion.

Figure 8. Imports of wind-powered generating sets, towers, and blades as a share of the U.S. market, 2005–09.

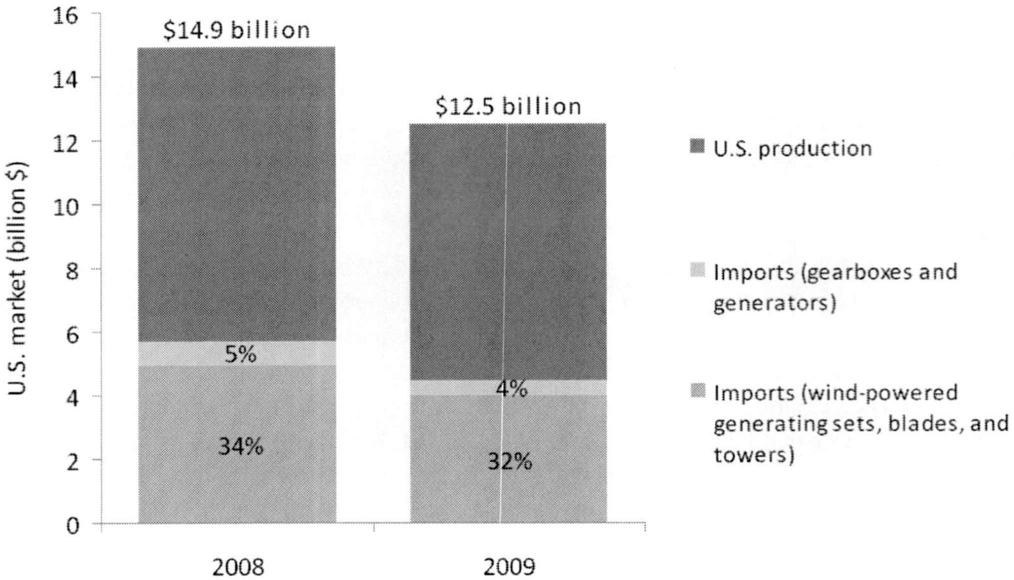

Sources: Import data for wind-powered generating sets, blades, and towers compiled from official statistics of the U.S. Department of Commerce; imports of gearboxes and generators are ITC calculations; value of the U.S. market based on USITC calculations using data on MW of wind turbines supplied provided by BTM Consult.

Notes: Value of the market calculated based on wind turbines supplied. Total value of wind turbines supplied includes both imports and U.S. wind turbine production. The figure does not include exports. Import data are based on the full value of imports of wind-powered generating sets (HTS 8502.31.0000) and towers and lattice masts (HTS 7308.20.0000) and two-thirds of the value of imports of parts of generators and other engines and motors (HTS 8503.00.9545 and 8412.90.9080). Import data are the landed duty-paid values. Imports of gearboxes and generators are ITC calculations based on estimates of U.S. production and the unit value of gearboxes and generators. See box 5 for a fuller discussion.

Figure 9. Imports of wind-powered generating sets, towers, blades, gearboxes, and generators as a share of the U.S. market, 2008–09.

Including two key nacelle components, gearboxes and generators, indicates a slightly higher import share and a slightly lower share of U.S. production.[52] Imports of gearboxes and generators accounted for a combined 5 percent of the value of the U.S. wind turbine market in 2008 and 4 percent in 2009 (figure 9).[53] In total, imports of wind-powered generating sets, blades, towers, gearboxes, and generators accounted for approximately 38 percent of the market in 2008 and 36 percent in 2009.[54] Thus even with the inclusion of two main components of the nacelle in this analysis, indications are that there is significant wind turbine production in the United States.

Box 5. Method for Estimating the Market Share of Imports

This paper estimates imports of wind-powered generating sets, blades, and towers as a share of the market from 2005 to 2009. Estimating imports as a share of the market is complicated by the difficulty in estimating the size of the market and by Harmonized Tariff Schedule (HTS) provisions that are not exclusively wind products.

Market Size Calculation

For this analysis, market size is estimated by multiplying the number of MWs supplied by each OEM by the average price per MW of wind turbines for that OEM. This calculation used BTM Consult data on wind turbines supplied rather than turbines installed to ensure that the results are not skewed by cases in which turbines are not imported and installed in the same year (see appendix).

Price per MW was calculated by dividing the sales or revenues of the OEM by MW sold in a given year. Where possible, service and other types of revenue were excluded from the calculation so that revenue reflects only equipment sales. To the extent possible, these data are based on actual sales and revenues as reported by producers. In one case, revenue and sales were not reported in annual reports, so data were calculated based on the best available information (e.g., public statements, presentations on quarterly earnings, etc.). In several other cases no data were available, so the price per MW was the average price per MW of five producers with sales in the U.S. market—Clipper, Gamesa, GE, Suzlon, and Vestas. For years before Clipper began producing in the United States, average prices for the other four companies were used. In most cases, price data were for global sales, not specifically for sales in the U.S. market. For producers who are not U.S.-based, revenue was converted to U.S. dollars by averaging each year's daily exchange rates using data from the Federal Reserve.

U.S. wind turbine market size, 2005–09

		2005	2006	2007	2008	2009
U.S. Market Size	Estimated turbine price (thousand $/MW)	1,093	1,097	1,199	1,432	1,450
	U.S. market size (thousand $)	2,694,119	3,292,111	7,904,476	14,876,025	12,490,453

Source: Price calculations by USITC staff; market size calculated based on BTM Consult data MW supplied.

Import Data and Use of HTS Provisions

The full value of the HTS provision for wind-powered generating sets was used in calculating the value of wind turbine imports since this provision is exclusive to wind. Some of the imports are small wind turbines, but this is generally a small percentage of imports.

USITC estimates of the percentage of HTS provisions that include towers and blades were used to determine the import share of wind turbines. The USITC estimates that from 2005 to 2009 towers accounted for the predominant share of imports in the provision for tower and lattice masts, so the full value of this provision was used. Blades can enter under one of two HTS provisions that are part of broad parts categories that can include components for a number of uses other than wind. Therefore, two-thirds of the value of imports in these provisions is used in calculating the market share of imports. The estimated import value of blades also includes certain other wind components such as hubs.

The import data used in this calculation are the landed duty-paid value. All other data in this paper is the customs value of imports. Landed duty-paid values are used here since it includes customs duties paid and cost, insurance, and freight and is likely the most reflective of the entire cost of an imported wind turbine.

Estimated imports of wind-powered generating sets, blades, and towers, 2005–09, thousand dollars

	2005	2006	2007	2008	2009
Imports	1,021,683	2,106,852	4,037,817	4,986,819	4,004,561

Source: Compiled from official statistics of the U.S. Department of Commerce.

Imports of Wind Turbine Gearboxes and Generators

Imports of gearboxes and generators in 2008 and 2009 were calculated based on estimates of nacelles assembled in the United States using imported gearboxes in 2008 and 2009 and unit values for wind turbine gearboxes and generators. Unit values are based on data on gearbox sales from Hansen Transmission and unit value of likely wind gearbox and generator imports from trade data. These are estimates of imports for new wind turbines and do not include replacement gearboxes and generators. Trade data were not used due to the difficulty in determining the share of the import category that is wind related.

Estimated imports of wind turbine gearboxes and generators, 2008–09, thousand dollars

	2008		2009	
	Unit Value/MW	Imports	Unit Value/MW	Imports
Gearboxes	123.2	389,000	116.2	244,000
Generators	52.6	323,000	49.6	214,000

Source: USITC calculation based on official statistics from the U.S. Department of Commerce and Hansen Transmissions, "Financial Year 2010 Results," May 17, 2010, 13.

Employment Growth and Wage Levels

The growth in the wind market has led to significant growth in wind-related employment. Employment in the wind industry, including manufacturing, construction, and service sector jobs, rapidly increased through late 2008, with total employment increasing from about 50,000 in 2007 to about 85,000 in 2008. About 20,000 of the 85,000 jobs were in manufacturing.[55] However, the financial crisis and the recession have had a negative impact on employment in the wind turbine manufacturing sector. In 2009, employment in the manufacturing sector declined by 1,500–2,000 jobs, though overall employment in the industry stayed at about 85,000 due to an increase in employment in construction and maintenance.[56]

Wages of production workers at wind industry manufacturing plants tend to be slightly above wages at similar plants. Based on 18 manufacturing plants for which information was publicly available, the average hourly wage of employees in wind industry manufacturing plants is generally between $13 and $20 (table 3). Comparing wages at these plants to national wages of production workers in the same industry, 61 percent of wind manufacturing plants pay above the national average wage for production occupations and 78 percent pay above the national median wage.[57] In addition, 61 percent pay above the average wage of all production workers, regardless of industry, in their state.[58]

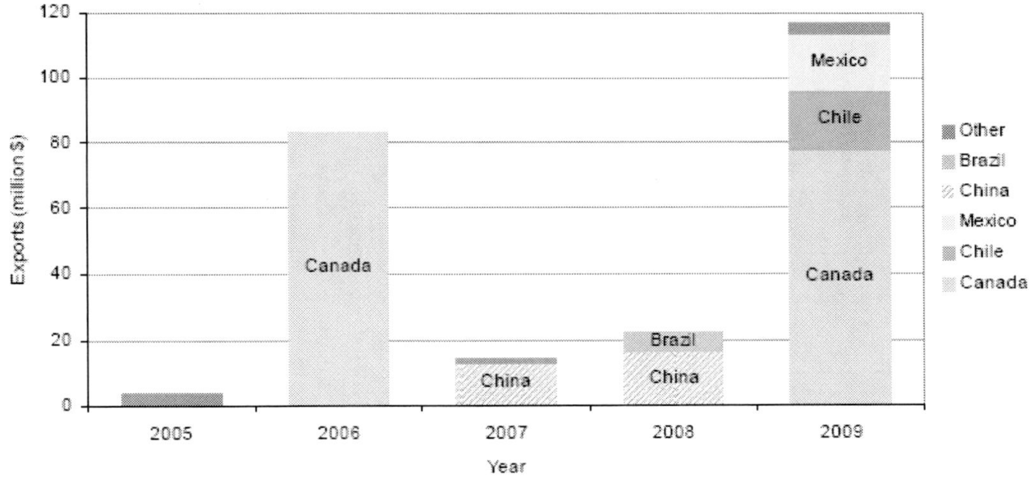

Source: Compiled from official statistics of the U.S. Department of Commerce.

Figure 10. U.S. exports of wind-powered generating sets, 2005–09.

Table 3. Estimated average hourly wages at 18 manufacturing plants compared to wages of production workers at the state and national level in 2008

	Number of wind plants in sample	Average wage range for wind plants (hourly)[a]	NAICS code[b]	National wages for production occupations, by NAICS[c]		Share of wind plants with wages above national wages, by NAICS[d]		Share of wind plants with wages above state average for production occupations[e] (percent)
				Average	Median	Above average wage (percent)	Above median wage (percent)	
Nacelle plants	6	$15 to $20	333611	$17.52	$16.49	33	67	100
Blade plants	7	$13 to $17	326199	$14.32	$13.09	57	71	29
Tower plants	5	$15 to $18	332312	$15.41	$14.47	100	100	60
All plants	18	$13 to $20				61	78	61

Sources: Plant wage data compiled from press releases; media reports; Kelly, "Wind Energy Update," June 26, 2008; and Mattera et al., *High Road or Low Road? Job Quality in the New Green Economy*, February 3, 2009, 15. National and state data on manufacturing wages are from Bureau of Labor Statistics (BLS), "May 2008 State Occupational Employment and Wage Estimates"'; BLS, "May 2008 National Industry- Specific Occupational Employment and Wage Estimates."

Note: Two of the nacelle plants also produce blades; these plants were only included in the nacelle category. The 18 plants are located in a total of nine states. Wage data are for plants for which information is publicly available.

[a] In several cases a wage range was given. In this case the median of the range was used as the average wage for that manufacturing plant. Most of the data are for new plants announced between 2006 and 2008, with most wage data from 2008. When possible, non- production workers are excluded from the sample. However, in several cases it was not clear whether production workers were included in the data.

[b] North American Industrial Classification System (NAICS) code.

[c] Wages are reported at 4-digit NAICS code level. The wage rate that is used is the rate for all production occupations within the NAICS code. 2008 data are used since this is when most of the manufacturing plants in the sample were announced.

[d] Figures in the column on the left were calculated by dividing the number of wind plants where the average wage is above the national average for that NAICS code by the total number of plants. Those in the column on the right were calculated by dividing the number of wind plants where the average wage is above the national median for that NAICS code by the total number of plants.

[e] Calculated by dividing the number of wind plants where the average wage is above the average wage for all production workers in the state by the total number of plants. State-level data are not available by NAICS code.

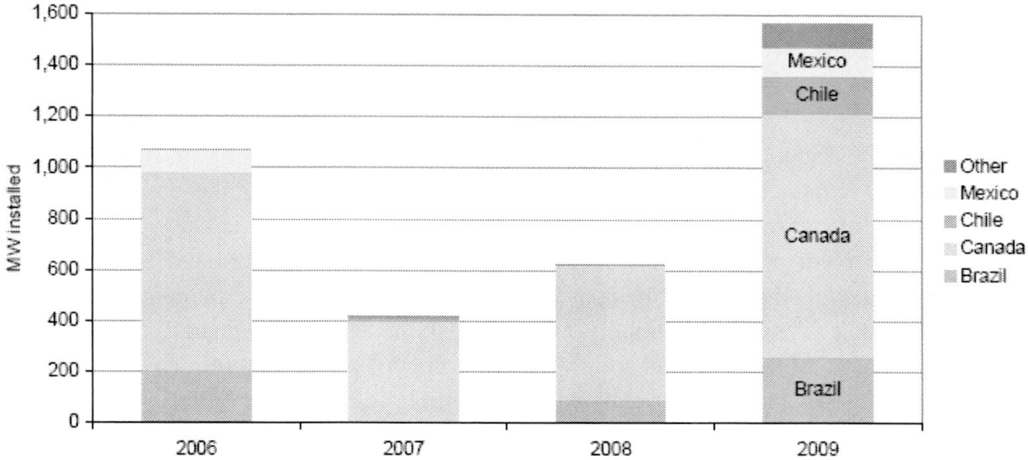

Source: Global Wind Energy Council (GWEC), *Global Wind 2008 Report*, April 2009; GWEC, "Global Wind Power Boom Continues," February 3, 2010.

Figure 11. Annual wind installations in the Americas, excluding the United States, 2006–09.

U.S. EXPORTS

The value of U.S. wind-powered generating set exports increased from $3.6 million in 2005 to $117.0 million in 2009 due to rising exports to Canada, Mexico, and South America (figure 1 0).[59] In 2009, Canada accounted for 66 percent of exports, Chile 16 percent, and Mexico 15 percent.[60] Overall, exports have been limited over the last five years due to the lack of U.S. production capacity, high demand within the U.S. market, and the fact that the largest U.S. producer, GE, has plants in many of the key foreign markets.[61]

Recent growth in U.S. production capacity and markets in Canada, Mexico, and South America indicate that there is the potential for further U.S. exports, but these exports might be limited by tariff and nontariff trade barriers and continued demand within the U.S. market. Installations in the Americas, excluding the United States, totaled 1,572 MW in 2009, and the market is expected to grow to 3,900 MW by 2013 (figure 11).[62] Canada and Brazil were the third and sixth largest importers, respectively, of wind- powered generating sets.[63] Some foreign OEMs and blade manufacturers have established manufacturing plants in the United States with the intention of serving both the U.S. market and other markets in the Americas.[64] However, tariff and nontariff barriers, increasing local production in Brazil and Canada, and the resumption of the growth of the U.S. market may limit U.S. exports.[65] Finally, the projected growth of North American markets to 3,900 MW would still leave these markets much smaller than the U.S. market.

CONCLUSION

U.S. production of wind turbines increased significantly from 2005 to 2008 and is likely to increase further as new production facilities come online in the next two years. Despite the appearance that imports, given the sharp increase from 2005 to 2008, now dominate the market, the opposite seems to be the case. The number of OEMs with at least one production plant in the United States increased from one in 2005 to nine in 2009. Another six have plants planned and there has been significant expansion in blade and tower manufacturing. The recession and the financial crisis were a challenge for the industry starting in the fall of 2008, but production remained at a much higher level than in 2005 and the construction and planning of additional manufacturing plants moved forward in 2009 and early 2010. Imports peaked at 64 percent of the market in 2006 and, though imports increased in value terms through 2008, they declined to 34 percent of the market in 2008 and 32 percent in 2009. It is difficult to predict whether U.S. imports will increase or decrease over the long term given the growing size of the U.S. market and the large number of new competitors seeking to enter the U.S. wind market. However, if planned manufacturing plants come online in the next few years, U.S. production capacity will continue to expand.

APPENDIX

Imports of wind-powered generating sets are generally related to installations, with imports preceding installations by a quarter (figure A. 1). Imports typically peak in the third quarter and installations in the fourth quarter, though fourth-quarter 2008 imports and first-quarter 2009 installations and imports were higher than usual. The change in import and installation patterns at the end of 2008 and in early 2009 makes it difficult to compare import and installation data in these years.

Imports of towers and lattice masts tend to precede installations and peak in the third quarter (figure A.2). Imports of parts of generators and other engines and motors also generally precede installations, though this HTS provision likely has the lowest wind content. In 2009, however, imports of towers and parts of generators and other engines and motors were flat in the first two quarters and then trended downward in the third and fourth quarters.

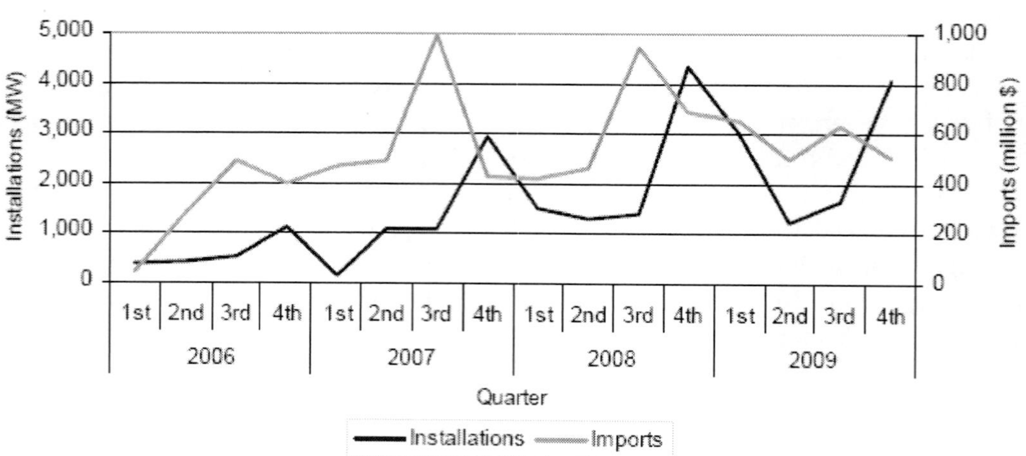

Source: Installation data from AWEA; trade data compiled from official statistics of the U.S. Department of Commerce.

Figure A.1. Wind turbine installations and wind-powered generating set imports, by quarter, 2006–09.

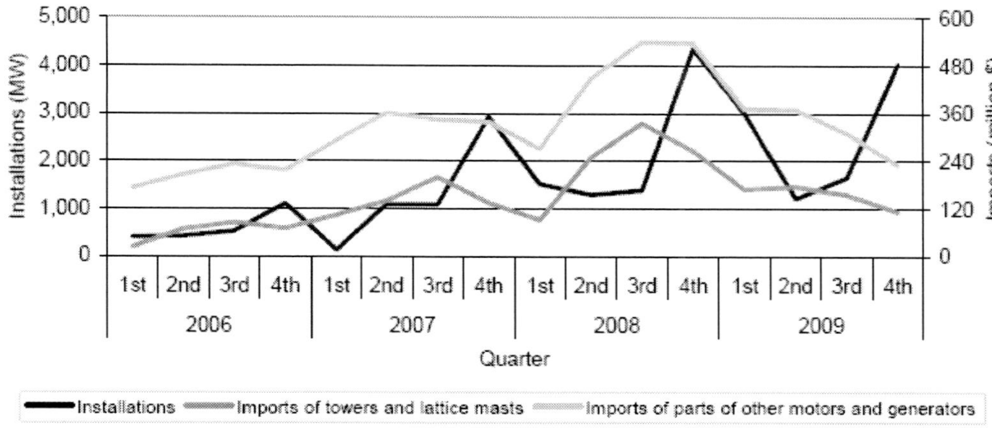

Source: Installation data from AWEA; trade data compiled from official statistics of the U.S. Department of Commerce.

Figure A.2. Wind turbine installations and imports of towers and lattice masts and parts of generators and other motors and engines, by quarter, 2006–09.

References

American Council on Renewable Energy. "*Overview: Renewable Energy Provisions; American Recovery and Reinvestment Act of 2009*," n.d. http://www.acore.org/files/images stimulus overview.pdf.

American Wind Energy Association. "Annual Rankings Demonstrate Continued Growth of Wind Energy in the United States." News release, March 15, 2006.

"Annual U.S. Wind Power Rankings Track Industry's Rapid Growth." News release, April 11, 2007.

———. *Annual Wind Industry Report: Year Ending 2008*, 2009. http://www.awea.org.

———. *2008 Annual Rankings Report*, April 2008. http://www.awea.org.

———.*Year End 2009 Market Report*, January 2010. http://www.awea.org.

American Wind Energy Association, Blue Green Alliance, and United Steelworkers. *Winds of Change: A Manufacturing Blueprint of the Wind Industry*, June 2010. http://www.awea.org.

Bauer, Julia. (2010). "Blades for Industrial-size Wind Turbines to Be Built in Holland at Former Yacht Facility." *Grand Rapids Press,* January 26. http://www.mlive.com/business/westmichigan/index.ssf/20 1 0/01/blades for industrial-size win.html.

Belyeu, Kathy. (2010). "Status of the U.S. Wind Power Market and Drivers for the Future." Presented at 2010 Windpower Conference, Dallas, TX, May 25.

Bloomberg New Energy Finance Database (accessed various dates).

BTM Consult ApS. *International Wind Energy Development: World Market Update 2008*, March 2009.

———. *International Wind Energy Development: World Market Update 2009*, March 2010.

Clipper Windpower. (2009). *Annual Report and Financial Statements for the Year Ended*

December 31, 2008, May. http://www.clipperwind.com.
———.*2009 Annual Report*, 2010. http://www.clipperwind.com.
———. "Operational and Trading Update." News release, March 9, 2010. http://www.clipperwind.com/pr 03092010 trading.php.
Del Franco, Mark. (2009). "Ontario Wind Comes under Microscope." *North American Windpower* 6, no. 8, September.
———. "States and the Supply Chain: More than a Trivial Pursuit." *North American Windpower* 7, no. 1, February 2010.
European Wind Energy Association. "Supply Chain: The Race to Meet Demand." *Wind Directions*, January/February 2007. http://www.ewea.org.
FDIMarkets Investment Database. http://www.fdimarkets.com (accessed November 3, 2008).
Franzman, Dave. (2009). "Update: Clipper Windpower Confirms Layoff at Cedar Rapids Plant." *KCRG*, January 20. http://www.kcrg.com/news/local/37883834.html.
Gamesa Corporación Tecnológica. (2008). "*3rd Quarter Results 2008 and Outlook Beyond 2010.*" Presentation, October 22. http://www.gamesacorp.com/files/Documentos%20PDF/Ingles/Informes%20Trimestrales/2008/P resentaci%F3n3%BAT%28 1%29.pdf.
General Electric Co. Transcript for Q4 2008 General Electric Earnings Conference Call, January 23, 2009. http://www.ge.com/pdf/investors Transcript 2009 01 23T 13 30.pdf.
———.Transcript for Q4 2009 General Electric Earnings Conference Call, January 22, 2010. http://www.ge.com/pdf/investors earnings transcript 01222010 english.pdf.
Global Trade Information Service, Inc. (GTIS). World Trade Atlas Database (accessed June 27, 2010).
Global Wind Energy Council. *Global Wind 2008 Report*, April 2009. http://www.gwec.net/index.php?id= 1 53&L=0%B4.
———. "Global Wind Power Boom Continues despite Economic Woes." News release, February 3, 2010.
Hall, Phil. (2010). "A Double Edged Sword: Examining Section 1603." *North American Windpower* 7, no. 1, February.
Hansen Transmissions. (2010). "*Financial Year 2010 Results." Analyst Meeting Presentation*, May 17. http://www.hansentransmissions.com/data/Hansen%20-%20Analyst%20meeting%20- %20%20FY%2020 1 0.pdf.
IHS Emerging Energy Research. "IHS Study: After Record-breaking 2009, US Wind Markets Face a Constrained 2010." News release, May 26, 2010.
Kaplan, Matthew. (2009). "*US Wind Growth Flat: Project Spillover Carries 2009.*" North American Wind Energy Advisory, Emerging Energy Research, October 23. http://www.emerging-energy.
Kelly, Marguerite. (2008). "Wind Energy Update." National Renewable Energy Lab, June 26. http://arkansasenergy.org/wind/awwg kelly.pdf.
Kirkegaard, Jacob Funk. Thilo, Hanemann. & Lutz, Weischer. (2009). *It Should Be a Breeze: Harnessing the Potential of Open Trade and Investment Flows in the Wind Energy Industry*. Working Paper 09– 14, Peterson Institute for International Economics and World Resources Institute, December 2009. http://www.iie.com/publications/wp/wp09-14.pdf.
Lewis, Joanna. & Ryan, Wiser. (2005). *Fostering a Renewable Energy Technology Industry:*

An International Comparison of Wind Industry Support Mechanisms. LBNL-59 116, Lawrence Berkeley National Lab, Environmental Energies Technology Division, November. http://eetd.lbl.gov/ea/emp/reports/59116.pdf.

Lillian, Jessica. (2009). "European Turbine Manufacturers Hot for U.S." *North American Windpower* 6, no. 4, May.

Lorinc, John. (2010). "Samsung Signs $6.6 Billion Deal to Build Wind and Solar Power in Ontario." *The New York Times*, January 21. http://greeninc.blogs billion-deal-to-build-wind-and-solar-power-in-ontario/.

Magee, Joshua. (2009). "*American Reinvestment and Recovery Act: Uncertainty in 2009 Paves the Way for Confidence in 2010.*" Presentation at Windpower 2009 Conference. Chicago, May 6. http://www.emerging-energy.

Mattera, Philip. Alec Dubro. Thomas, Gradel. Rachel, Thompson. Kate, Gordon. & Elena Foshay. (2009). *High Road or Low Road? Job Quality in the New Green Economy*. Good Jobs First, February 3. http://www.goodjobsfirst.org/pdf/gjfgreenjobsrpt.pdf.

Mellott, Kathy. (2009). "Gamesa Layoffs a Surprise." *The Johnstown (PA) Tribune-Democrat*, November 27. http://tribune-democrat.com/local/x546432165/Gamesa-layoffs

Norfleet, Gregory R. (2009). "Acciona Expects to Rebound Soon after Layoffs." *West Branch (IA) Times*, March 25. http://www.westbranchtimes.com/article.php?viewID=4186.

Pettinger, Frank. (2009). "Continental Wind Energy Targets Medium Market." *Reuters*, August 11. http://www.reuters.com.

Pioneer Power Solutions. "*Pioneer Power Solutions Completes Asset Purchase from AAER Inc.*" News release, June 7, 2010.

Port of Longview. "*Port of Longview Handles Blades for Export*." News release, August 14, 2008.

Recharge. "Daewoo, Nova Scotia Sign Wind Turbine Joint Venture." *Recharge*, March 8, 2010. http://www.rechargenews.com/business area/finance/article208221 .ece.

Reuters. "Chinese Eye Domination of Wind Turbine Market." *Reuters*, May 26, 2010. http://www.reuters.com/article/idUSTRE64P26920100526.

Richards, Timothy J. (2009). Written testimony submitted to the House Committee on Energy and Commerce, Subcommittee on Commerce, Trade, and Consumer Protection, hearing on Growing Trade in Green Technology, October 7. http://energycommerce. house.gov/Press 111/20091007/richards testimony.pdf.

Sala de Vedruna, Eduard. (2010). "*New Wind Capacity Surges in 2009, Despite Economy*." Global Wind Energy Advisory, IHS Emerging Energy Research, March 31. http://www.emerging-energy.

Schmit, Julie. (2010). "Wind-energy Industry Lost Factory Jobs Despite Stimulus." *USA Today*, January 26. http://www.usatoday.com/money/industries N.htm.

Siegel, Jeff. (2010). "Armed with Cash, China Shops in U.S. Wind Market." *North American Windpower* 7, no. 5, June.

U.S. Department of Labor. Bureau of Labor Statistics. "May 2008 State Occupational Employment and Wage Estimates." http://www.bls.gov/.

———. "May 2008 National Industry-Specific Occupational Employment and Wage Estimates." http://www.bls.gov/.

U.S. International Trade Commission (USITC). (2009). *Industry and Trade Summary: Wind Turbines*. USITC Publication ITS–02. Washington, DC: USITC, June.

http://www.usitc.gov/publications/332/ITS-2.pdf.

———.Interactive Tariff and Trade Dataweb (Dataweb). http://dataweb.usitc.gov (accessed various dates).

Vestas. *Annual Report 2009,* n.d. http://www.vestas.com/en/investor.aspx.

———. "First Train from Pueblo." News release, May 31, 2010.

White House. (2010). "*Fact Sheet: $2.3 Billion in New Clean Energy Manufacturing Tax Credits*." News release, January 8.

Windpower Monthly. (2008). "A Market Ready for the Next Big Step." *Windpower Monthly Special Report: A Guide to Windpower Investment in Canada*, October.

———. "Nordex Delays U.S. Factory Debut," *Windpower Monthly*, May 1, 2009. Available for a fee at http://www.windpowermonthly.com/news/login/951352/.

End Notes

[1] For the purposes of this paper, "domestic production" refers to wind turbine manufacturing/assembly and does not refer to wind energy generation from wind farms. The terms U.S. production and domestic production include production at all plants in the United States regardless of whether they are owned by U.S.-based or foreign-based companies.

[2] Small wind turbines are produced by different companies for a different market.

[3] Unless otherwise noted, data on annual wind turbine installations cited in this report are net capacity: megawatts of wind capacity added, minus megawatts of wind capacity from turbines that are decommissioned. However, the capacity decommissioned each year in the United States and other countries is generally small.

[4] Global Wind Energy Council (GWEC), *Global Wind 2008 Report*, April 2009, 10; GWEC, "Global Wind Power Boom Continues," February 3, 2010.

[5] GWEC, "Global Wind Power Boom Continues," February 3, 2010.

[6] OEMs are companies that manufacture wind turbines and sell the turbines under their name (e.g., Acciona, Clipper, GE, Siemens, and Vestas). Wind project developers generally contract with OEMs for the delivery of the entire turbine, including the nacelle, blades, and tower. Most OEMs assemble the nacelle in-house, while blades, towers, and other components are produced either by the OEM or to the OEM's specifications by a supplier. This paper focuses on wind turbine manufacturing and does not discuss companies such as AMSC that develop and license wind turbine designs but do not manufacture nacelles in-house.

[7] For more on the factors that led to the development of the wind industry in Europe, see Lewis and Wiser, "Fostering a Renewable Energy Technology Industry," November 2005.

[8] Data on wind turbines supplied from BTM Consult, *International Wind Energy Development Update 2009*, March 2010, 28, 31, 33, 93, 113.

[9] Siegel, "Armed with Cash," June 2010, 80–83; *Reuters*, "Chinese Eye Domination of Wind Turbine Market," May 26, 2010; industry official, interview by Commission staff, Dallas, TX, May 24, 2010; industry official, interview by Commission staff, Dallas, TX, May 25, 2010.

[10] AWEA, *Annual Wind Industry Report: Year Ending 2008*, 2009, 4; AWEA, *Year End 2009 Market Report*, January 2010, 1–2.

[11] In 2009 first-quarter installations were 3,006 MW, second-quarter installations were 1,218 MW, third-quarter installations were 1,656 MW, and fourth-quarter installations were 4,041 MW. AWEA, *Year End 2009 Market Report*, 5, 9, 11, 13. For a general discussion of factors affecting wind turbine demand, see U.S. International Trade Commission (USITC), *Industry and Trade Summary: Wind Turbines*, June 2009, 25–30. For an overview of renewable energy provisions in the Stimulus Bill, see American Council on Renewable Energy, "Overview, Renewable Energy Provisions," n.d. For more on the impact of the Stimulus Bill, the U.S. market, and policy drivers in 2010, see, for example, Hall, "A Double Edged Sword: Examining Section 1603," February 2010, 18–22; Kaplan, "U.S. Wind Growth Flat," 2009; Magee, "American Reinvestment and Recovery Act," 2009; Sala de Vedruna, "New Wind Capacity," 2010.

[12] BTM Consult, *International Wind Energy Development Update 2009*, 48; IHS Emerging Energy Research, "IHS Study: After Record-Breaking 2009," May 26, 2010.

[13] From November 2009 to March 2010, there were only three publicly announced wind turbine orders. Magee, "American Reinvestment and Recovery Act," 2009.

[14] Clipper Windpower, "Operational and Trading Update," 2010; Norfleet, "Acciona Expects to Rebound Soon after Layoffs," 2009; Vestas, *Annual Report 2009*, n.d., 22.

[15] AWEA, *Year End 2009 Market Report*, 5, 9, 11, 13; Hall, "A Double Edged Sword: Examining Section 1603," 2010, 18–22; Sala de Vedruna, "New Wind Capacity," 2010.

[16] Industry official, interview by Commission staff, Washington, DC, February 24, 2010; Vestas, *Annual Report 2009*, 18; Lillian, "European Turbine Manufacturers Hot for U.S.," 21.

[17] Lillian, 20.

[18] AWEA, *Annual Wind Industry Report: Year Ending 2008*, 10; Belyeu, "Status of the U.S. Wind Power Market," May 25, 2010.

[19] While this is global deliveries for GE, it likely reflects trends in the U.S. market, since more than 80 percent of MW supplied by GE in 2008 and 2009 were to the United States. BTM Consult, *International Wind Energy Development Update 2008*, 24, 109; BTM Consult, *International Wind Energy Development Update 2009*, 28, 113; General Electric Co., "Conference Call Transcript for Q4 2009," 2010, 16; General Electric Co., "Conference Call Transcript for Q4 2008," 2009, 12.

[20] Clipper Windpower, *2009 Annual Report*, 2010, 38; Clipper Windpower, *Annual Report and Financial Statements*, May 2009, 3; AWEA, *Annual Wind Industry Report: Year Ending 2008*, 10; Belyeu, "Status of the U.S. Wind Power Market," May 25, 2010.

[21] The Danish company Vestas produced blades in the United States in 2009, but not nacelles. Vestas, *Annual Report 2009*, 26; AWEA, *Annual Wind Industry Report: Year Ending 2008*, 10; Belyeu, "Status of the U.S. Wind Power Market," May 25, 2010.

[22] Kaplan, "U.S. Wind Growth Flat."

[23] Compiled from press releases and media reports. For example, see Norfleet, "Acciona Expects to Rebound Soon after Layoffs," 2009; Franzman, "Update: Clipper Windpower Confirms Layoff at Cedar Rapids Plant," 2009; Mellott, "Gamesa Layoffs a Surprise," 2009; *Windpower Monthly,* "Nordex Delays U.S. Factory Debut," 2009.

[24] There were nine OEMs producing nacelles, blades, and/or towers in the United States in 2009 compared with only one in 2005. The one OEM with U.S. nacelle production, GE, accounted for 1,433 MW in installations in 2005. In 2009, companies with U.S. production accounted for 5,410 MW in installations. Most of the nacelles for these installations were likely produced in the United States. Only one OEM officially put its planned manufacturing plant on hold, though there is at least one other plant whose status is uncertain. Most planned plants moved forward, although often at a slower pace than initially anticipated. Information on OEMs producing in the United States and production data are compiled from press releases, media reports, and company documents and Web sites; data on wind turbine installations by U.S. manufacturers are USITC calculations based on AWEA data.

[25] USITC calculation based on Belyeu, "Status of the U.S. Wind Power Market," May 25, 2010.

[26] GE also has plants in Germany, Norway, China, Canada and the United States and is planning to start assembling nacelles in Brazil. GE Energy Web site, http://www.gepower.com; Bloomberg New Energy Finance database (accessed March 22, 2010).

[27] BTM Consult, *International Wind Energy Development Update 2009*, 28.

[28] Other companies, such as Continental Wind Power, have announced plans to begin U.S. production of utility-scale wind turbines, but this discussion is focused on companies with prior U.S. or global production of small or large wind turbines. Pettinger, "Continental Wind Energy Targets Medium Market," 2009; Northern Power Systems Web site, http://www.northernpower.com (accessed September 15, 2009).

[29] Compiled from press releases, media reports, company documents and Web sites, AWEA publications, and White House Web site (for the American Reinvestment and Recovery Act of 2009 [ARRA] Advanced Energy Manufacturing Tax Credit [MTC] list of awardees), http://www.whitehouse.gov/the-press-office/fact-sheet-23-billion-new-clean-energy-

[30] Industry official, interview by Commission staff, October 30, 2008; industry official, telephone interview by Commission staff, December 10, 2008.

[31] Belyeu, "Status of the U.S. Wind Power Market," May 25, 2010.

[32] Compiled from press releases, media reports, AWEA publications, and company Web sites.

[33] TPI Composites was also an awardee for a new manufacturing plant, though a site has not yet been announced. Bauer, "Blades for Industrial-size Wind Turbines," 2010; White House Web site (for the list of ARRA MTC awardees).

[34] Compiled from press releases, media reports and AWEA publications.

[35] Compiled from press releases, media reports, and the White House web site (for the list of ARRA MTC awardees).

[36] For list of current and planned manufacturing plants, as of June 2009, see USITC, *Industry and Trade Summary: Wind Turbines*, 2009, 13.

[37] AWEA, Blue Green Alliance, and United Steelworkers, *Winds of Change*, June 2010, 14.

[38] See the White House Web site (for the ARRA MTC list of awardees).

[39] For more on the percent of components that OEMs produce in-house, see USITC, *Industry and Trade Summary: Wind Turbines*, 2009, 10–12.

[40] Compiled from press releases, media reports, and company documents and Web sites.

[41] Revenues of Ameron International, DMI (part of Otter Tail Corporation), Tower Tech (Broadwind), and Trinity Structural Towers (Trinity Industries). In all, these tower manufacturers have nine combined U.S. factories, one factory in Canada, and one factory in Mexico. Since separate data are not available, figures include revenue from production at the facilities in Canada and Mexico. These companies likely represent a significant share of U.S. tower production in 2009, though precise data are not available. Data are taken from company statements and annual reports, but in some cases are the result of ITC calculations based on data included in these documents. The Ameron fiscal year ends on November 30, but fiscal-year data were used to represent the calendar year.

[42] Compiled from press releases, media reports, FDIMarkets investment database (accessed November 3, 2008), and Bloomberg New Energy Finance database (accessed March 12, 2010). Investment is listed under the year in which the plant was announced. In one case the amount of the manufacturing investment was not available. In this case, the amount invested by another company in a similar size manufacturing plant was used to approximate the investment amount. For joint ventures, half of the total investment is included.

[43] Wind-powered generating sets include the nacelle and, if imported with the nacelle, components such as blades and hubs. Blades, towers, and other components are included in separate provisions when they are not imported with the nacelle.

[44] Compiled from official statistics of the U.S. Department of Commerce.

[45] Blades and other components such as hubs, when imported separately from the nacelle, are included in HTS 8412.90.9080 (parts of other engines and motors) and HTS 8503.00.9545 (parts of generators).

[46] Compiled from official statistics of the U.S. Department of Commerce.

[47] Towers are included in HTS 7308.20.0000, towers and lattice masts.

[48] The reason for this decline in imports from Canada and Mexico is not clear, but it may not be indicative of a long-term trend. In the first four months of 2010, imports of towers and lattice masts from Canada totaled $31 million and from Mexico $40 million. Tower production in both countries is expanding, with Korean companies in particular investing in new plants in Canada and Mexico (though the extent to which the new plants in Canada will export to the United States is not clear). Trade data compiled from official statistics of the U.S. Department of Commerce; information on new tower plants compiled from press releases and media reports.

[49] BTM Consult collects data on the number of turbines supplied by manufacturers to the U.S. market. This may differ from the number of turbines installed because of a possible lag between when the turbine is shipped and when it is installed at a wind farm. Turbines supplied is the better metric for market size for this analysis since it is more closely correlated with imports. This is particularly true for 2008 and 2009 since, as noted earlier, 1.5 GW were shipped in 2008 but not installed until 2009.

[50] This does not control for the change in wind turbine values over time. Calculated by dividing the estimated import value of wind-powered generating sets, blades, and towers by the number of MW supplied. The value of imports is based on the full value of imports of wind-powered generating sets (HTS 8502.31.0000) and towers and lattice masts (HTS 73 08.20.0000) and two-thirds of the value of imports of parts of generators and other engines and motors (HTS 8503.00.9545 and 8412.90.9080). Import values are the customs values. Data on wind turbines supplied from BTM Consult, *International Wind Energy Development Update 2008*, 109; BTM Consult, *International Wind Energy Development Update 2009*, 113; industry official, e-mail message to Commission staff, September 18, 2009; trade data compiled from official statistics of the U.S. Department of Commerce.

[51] Calculations based on supply data from BTM Consult, *International Wind Energy Development Update 2008*, 109; BTM Consult, *International Wind Energy Development Update 2009*, 113; industry official, e-mail message to Commission staff, September 18, 2009; trade data compiled from official statistics of the U.S. Department of Commerce. See box 5 for an explanation of the methodology.

[52] Wind turbine gearboxes and generators combined account for approximately 15 to 20 percent of the cost of a wind turbine. European Wind Energy Association, "Supply Chain: The Race to Meet Demand," January/February 2007, 33; Gamesa, "3rd Quarter Results 2008 and Outlook Beyond 2010," 2008, 33.

[53] Imports of gearboxes and generators in 2008 and 2009 were calculated based on estimates of nacelles assembled in the United States using imported gearboxes in 2008 and 2009 and unit values of wind turbine gearboxes and generators. Data for earlier years were not included because of the challenges in estimating U.S. demand for gearboxes and generators. See box 5 for a fuller explanation.

[54] Numbers may not add due to rounding.

[55] AWEA, *Annual Wind Industry Report: Year Ending 2008*, 17.

[56] Schmit, "Wind-energy Industry Lost Factory Jobs Despite Stimulus," January 26, 2010.

[57] Wages are reported at the 4-digit North American Industrial Classification System (NAICS) code level. The wage rate that is used is the rate for all production occupations within the NAICS code.

[58] Plant wage data compiled from press releases; media reports; Kelly, "Wind Energy Update," June 26, 2008; Mattera et al., *High Road or Low Road? Job Quality in the New Green Economy*, February 3, 2009, 15. National and state data on manufacturing wages are from Bureau of Labor Statistics (BLS), "May 2008 State

Occupational Employment and Wage Estimates"; BLS, "May 2008 National Industry-Specific Occupational Employment and Wage Estimates."

[59] Similar export data for blades and towers is not included since it is difficult to determine how much of these export provisions are wind related. There is anecdotal evidence of exports of towers (e.g., tower exports to Canada and Mexico) and blades (e.g., blade exports to Australia), but information on the overall value of trade is not available. Industry official, interview by Commission staff, Dallas, TX, May 23, 2010; Port of Longview, "Port of Longview Handles Blades for Export," August 14, 2008; Vestas, "First Train from Pueblo," May 31, 2010.

[60] Compiled from official statistics of the U.S. Department of Commerce.

[61] GE has plants in Germany, Norway, China, Canada and the United States and is planning to start assembling wind turbines in Brazil. GE Energy Web site, http://www.gepower.com; Bloomberg New Energy Finance database (accessed March 22, 2010).

[62] BTM Consult, *International Wind Energy Development Update 2009*, 48; GWEC, "Global Wind Power Boom Continues despite Economic Woes," 2010.

[63] Global Trade Information Service, Inc. (GTIS), World Trade Atlas database (accessed June 27, 2010).

[64] Industry official, telephone interview by Commission staff, December 10, 2008; press releases and media reports.

[65] For example, at least two Canadian provinces—Ontario and Quebec—have local-content requirements, and Brazil raised duties on wind turbines of up to 2.6 MW to 14 percent in 2009. Companies with current nacelle manufacturing plants in Canada include Pioneer Wind (as a result of its purchase of AAER assets) and GE. In January 2010 Samsung reached an agreement with the government of Ontario to invest in wind and solar projects and related solar and wind equipment manufacturing. DSME announced plans in February 2009 to build a blade and tower plant in Canada, joining several other companies already producing towers and at least one other company producing blades in Canada. Two companies, Enercon and IMPSA, currently have nacelle facilities in Brazil, and since August 2009 at least four more OEMs—Alstom, Fuhrländer, GE, and Suzlon—have announced plans to assemble nacelles in Brazil. Enercon and Tecsis currently produce blades in Brazil, and both blade and tower production are expected to expand. *Windpower Monthly*, "A Market Ready for the Next Big Step," October 2008, 5–6; Del Franco, "States and the Supply Chain: More than a Trivial Pursuit," February 2010, 27; Del Franco, "Ontario Wind Comes under Microscope," September 2009, 42–43; Lorinc, "Samsung Signs $6.6 Billion Deal to Build Wind and Solar Power in Ontario," January 21, 2010; *Recharge*, "Daewoo, Nova Scotia Sign Wind Turbine Joint Venture," March 8, 2010; Bloomberg New Energy Finance database (accessed March 22, 2010); Marmen Web site, http://www.marmen.qc.ca; Hitachi Web site, http://hitachi.sasktelwebhosting.com; Kirkegaard, Hanemann, and Weischer, "It Should Be a Breeze," December 2009, 20; Richards, House Commerce, Trade, and Consumer Protection Subcommittee written testimony, October 7, 2009, 6; Pioneer Power Solutions, "Pioneer Power Solutions Completes Asset Purchase from AAER Inc.," June 7, 2010.

In: U.S. Wind Power Industry Market Conditions and Trends ISBN: 978-1-61209-559-2
Editor: Liam E. Dempsey

Chapter 3

WIND TURBINES: INDUSTRY AND TRADE SUMMARY

U.S. International Trade Commission

ABSTRACT

This report addresses trade and industry conditions for wind turbine manufacturing for the period 2003 to 2008.

- *Annual U.S. wind turbine installations* increased from 1,672 to 8,545 megawatts (MW) from 2003 to 2008 due to the growing cost competitiveness of wind energy, advances in wind technology, and government policies. The United States was the global leader in installations in 2008, with 31 percent of global wind turbine installations. China (6,300 MW, 23 percent of installations), India (1,800 MW, 7 percent), Germany (1,665 MW, 6 percent), and Spain (1,609 MW, 6 percent) were second through fifth, respectively, in installed wind capacity. Installations represent the size of the market for wind turbines.
- *Domestic production capacity* increased due to the growth of the U.S. wind energy market from 2003 to 2008. Five companies (Acciona, Clipper, CTC/DeWind, Gamesa, and GE) currently manufacture nacelles in the United States (the nacelle houses the generator and gearbox). Six more companies are planning to open nacelle plants in 2009 or 2010. There are at least 10 blade and 15 tower manufacturers with plants open or planned.
- *Manufacturing employment* increased to about 20,000 in 2008. Hourly wages in wind turbine plants generally average between $13 and $20. Some manufacturers announced layoffs in late 2008 and early 2009 due to the financial crisis.
- *Competition* in the U.S. market increased. The number of turbine manufacturers with U.S. sales increased from five in 2003 to 13 in 2008. U.S.-based General Electric (GE) was the leading manufacturer in the U.S. market in 2008 with 43 percent of the market, followed by Danish manufacturer Vestas with 13 percent, and German manufacturer Siemens with 9 percent. U.S.-based Clipper Windpower began production in 2006 and had a 2008 market share of 7 percent.
- *Factors that affect demand* for wind turbines include energy prices, availability of project financing, and government policies such as renewable portfolio standards and the production tax credit. Insufficient transmission capacity, difficulties in obtaining permits to transport turbines, periodic expirations of the production tax credit, and

financing constraints due to the financial crisis may be impediments to future U.S. wind energy development. The American Recovery and Reinvestment Act mitigates some of these constraints by expanding and extending tax credits, making wind eligible for additional financing, and investing in transmission.

- *The United States is a net importer* of wind turbines and major components. *U.S. imports* of wind- powered generating sets increased from $356 million in 2003 to $2.5 billion in 2008 (more than 600 percent) while *U.S. exports* increased from $0.7 million in 2003 to $22.1 million in 2008. The U.S. trade deficit in 2008 for wind turbines was $2.5 billion. Denmark and Spain accounted for a combined 55 percent of imports in 2008. The largest U.S. export markets were China (73 percent of exports) and Brazil (27 percent).
- *The United States trails Europe and Asia* in the manufacturing and export of wind turbines and components. Denmark, Germany, India, Japan, and Spain accounted for a combined 91 percent of global exports of wind-powered generating sets in 2008.

ACRONYMS

ACORE	American Council on Renewable Energy
AEE	Spanish Wind Energy Association
AWEA	American Wind Energy Association
CTC	Composite Technology Corporation
DOE	Department of Energy
DSIRE	Database of State Incentives for Renewables and Energy Efficiency
EERE	DOE Office of Energy Efficiency and Renewable Energy
EIA	Energy Information Administration
EWEA	European Wind Energy Association
EWT	Emergya Wind Technologies
GW	Gigawatt
GWEC	Global Wind Energy Council
IEA	International Energy Agency
IPP	Independent Power Producer
ITC	Investment Tax Credit
kW	Kilowatt
kWh	Kilowatt hour
MW	Megawatt
MWh	Megawatt hour
NERC	National Electric Reliability Council
NREL	National Renewable Energy Lab
PTC	Production Tax Credit
RPS	Renewable Portfolio Standard
R&D	Research and Development
RD&D	Research, Development, and Demonstration

INTRODUCTION

U.S. wind turbine manufacturing and trade are rapidly expanding due to growing domestic demand for wind turbines. Annual U.S. wind turbine installations[1] increased from 1,672 to 8,545 megawatts (MW) (over 400 percent) from 2003 to 2008, leading to a growth in imports and rising investment in U.S. production by American, European, and Asian companies.[2] New U.S. manufacturing plants are primarily intended to meet domestic demand, but some U.S. manufacturers are beginning to examine export opportunities or open overseas manufacturing plants.

This Summary analyzes utility scale wind turbine[3] manufacturing, investment, and trade from 2003 to 2008.[4] The first section analyzes the U.S. manufacturing industry, domestic employment, and the supply chain. The second section examines competition in the U.S. market, turbine demand, and government policy. The final sections analyze U.S. import and export trends, foreign industry, foreign markets, and global trade treatment.

The growing importance of the wind industry is reflected in U.S. trade flows and U.S. manufacturing production. From 2003 to 2008, imports of wind-powered generating sets increased more than 600 percent to $2.5 billion annually, while domestic wind turbine sales of the largest U.S. manufacturer, GE, rose by over 300 percent from 2003 to 2008.[5] The number of competitors in the U.S. market increased from five in 2003 to 13 in 2008.[6] At least 11 manufacturers now have nacelle (box 1) assembly plants open or planned in the United States, only one of which had domestic manufacturing capabilities in 2003. There has been a similar expansion in production of blades and towers.

U.S. exports are limited, but there are some indications that the presence of U.S. companies in overseas markets is growing. U.S. exports of wind-powered generating sets increased from $0.7 million in 2003 to $22.1 million in 2008.[7] Companies are establishing U.S. manufacturing plants with the intention of serving markets throughout North and South America and some U.S. companies have invested in foreign production in order to enter overseas markets.

The financial crisis has led to a decrease in demand in the short term, but long-term demand will likely rebound due government policies that support renewable energy, the recovery of the credit markets, and the return to the market of tax equity investors. There are several impediments to further wide scale deployment of wind turbines, but manufacturers are moving ahead with planned investments in U.S. production in the expectation that the market will grow in the long term.

BOX 1. WIND TURBINE COMPONENTS

Blades: Most wind turbines have three blades, though there are some with two blades. Blades are generally 30 to 50 meters (100 to 165 feet) long, with the most common sizes around 40 meters (130 feet). Longer blades are being designed and tested. Blade weights vary, depending on the design and materials—a 40 meter LM Glasfiber blade for a 1.5 MW turbine weighs 5,780 kg (6.4 tons) and one for a 2.0 MW turbine weighs 6,290 kg (6.9 tons).

Controller: There is a controller in the nacelle and one at the base of the turbine. The controller monitors the condition of the turbine and controls the turbine movement.

Gearbox: Many wind turbines have a gearbox that increases the rotational speed of the shaft. A low-speed shaft feeds into the gearbox and a high-speed shaft feeds from the gearbox into the generator. Some turbines use direct drive generators that are capable of producing electricity at a lower rotational speed. These turbines do not require a gearbox.

Generators: Wind turbines typically have a single AC generator that converts the mechanical energy from the wind turbine's rotation into electrical energy. Clipper Windpower uses a different design that features four DC generators.

Photos courtesy of DOE/N REL, Credit – Lee Fingersh.

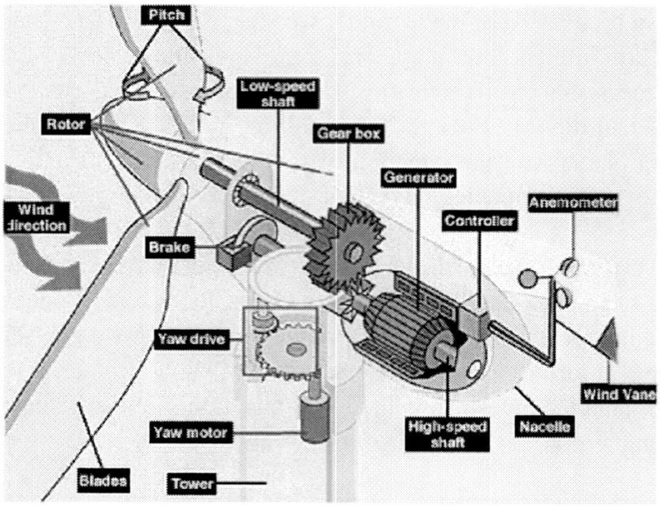

Source: DOE, EERE Web site. http://www.eere.energy (accessed April 1, 2009).

Nacelles: The nacelle houses the main components of the wind turbine, such as the controller, gearbox, generator, and shafts.

Rotor: The rotor includes both the blades and the hub (the component to which the blades are attached).

Towers: Towers are usually tubular steel towers 60 to 80 meters (about 195 to 260 feet) high that consist of three sections of varying heights. (There are some towers with heights around 100 meters (330 feet)).

Sources: Danish Wind Industry Association Web site. http://www.windpower.org (accessed various dates); industry official, interview by Commission staff, October 30, 2008; LM Glasfiber Web site. http://www.lmglasfiber.com (accessed March 7, 2009); and U.S. Department of Energy (DOE), Office of Energy Efficiency and Renewable Energy (EERE) Web site. http://www.eere.energy (accessed April 1, 2009).

INDUSTRY PROFILE

The U.S. wind turbine manufacturing industry is rapidly expanding due to the increase in wind turbine installations in the United States. The number of original equipment manufacturers (OEMs) (box 2) assembling nacelles in the United States increased from one in 2003 to five in 2008 (table 1).[8] Six more OEMs are expected to open plants in 2009 or 2010. OEMs currently produce turbines in the United States that range between 1.5 and 2.5 megawatts (MW).[9] The range of sizes produced in the United States will expand in 2009 and 2010. Acciona Windpower plans to begin production of a 3.0 MW turbine in 2009 and two of the new facilities will produce turbines of 1 MW or less.[10]

BOX 2. GLOBAL ORIGINAL EQUIPMENT MANUFACTURERS

The global wind turbine manufacturing industry is dominated by a small number of OEMs. The top five OEMs, Vestas, GE, Gamesa, Enercon, and Suzlon, had a combined 62 percent of the global market in 2008. GE is the leading U.S.-based supplier and was second in the world in market share in 2008. Most other companies are European and Asian OEMs. Many European OEMs are among the most mature and competitive global companies. Manufacturing capacity in Asia is growing rapidly and five Asian companies are now among the 13 largest OEMs.

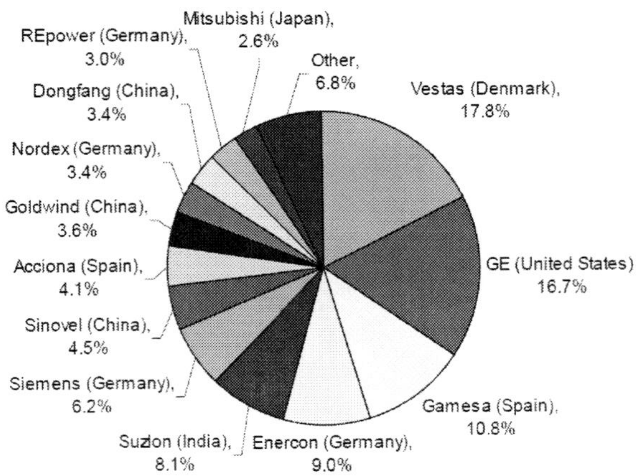

Source: BTM Consult, *International Wind Energy Development: World Market Update 2008*, 24.

Table 1. U.S. nacelle manufacturing plants, operational and planned, 2009

Company	Plant Location(s)	Turbine Size (MW)	Production Capacity (Turbines)	Status	Headquarters
Acciona Windpower	Iowa	1.5 and 3.0	400	Operational	Spain
Clipper Windpower	Iowa	2.5	More than 400	Operational	United States
CTC/DeWind	Texas	2.0	500	Operational	United States
Emergya Wind Technologies	Arkansas	750 or 900 kW	NA	Expected to begin production in 2009	Netherlands
Fuhrlander	Montana	2.5	NA	Expected to begin production in 2009 or 2010	Germany
Gamesa	Pennsylvania	2.0	500	Operational	Spain
General Electric	California, Florida, South Carolina	1.5 and 2.5	NA	Operational	United States
Nordex	Arkansas	2.5	300	Expected to begin production in 2010	Germany
Nordic Windpower	Idaho	1.0	240	Expected to begin production in 2009	United Kingdom
Siemens	Kansas	2.3	650	Expected to begin production in 2010	Germany
Vestas Wind Systems	Colorado	NA	1,400	Expected to begin production in 2010	Denmark

Sources: Compiled from AWEA publications, media reports, press releases, and interviews.

Notes: NA: Not available. Acciona is expected to begin production of the 3.0 MW turbine in Iowa in 2009 and GE is expected to introduce the 2.5 MW turbine to the U.S. market in 2010. Some of the manufacturing plants may also produce blades and other components. Emergya Wind Technologies (EWT) nacelles will be assembled in Arkansas by Wind & Water Technology. Blades will be manufactured at the same facility by EWT subsidiary Polymarin Composites. CTC/DeWind turbines are manufactured by TECO Westinghouse Motor Company in Texas.

Domestic OEMs are expanding production or entering the industry for the first time. General Electric Co. (GE) is the largest company in the U.S. market, with 43 percent of the market[11] in 2008 and nacelle assembly plants (boxes 3 and 4) in California, Florida, and South Carolina.[12] U.S.-based Clipper Windpower produced its first wind turbines in 2006 and had 7 percent of the U.S. market in 2008.[13] Composite Technology Corporation (CTC) purchased European wind turbine manufacturer EU Energy, now DeWind, in 2006 and subsequently established manufacturing operations in the United States.[14] Other U.S. companies are considering entering the industry. For example, Vermont based Northern

Power Systems, which currently produces a 100 kW wind turbine, is developing a 2.2 MW turbine for the U.S. market.[15]

Box 3. Production Methods

OEMs typically have unique designs for wind turbine nacelles and the leading global OEMs manufacture nacelles in-house. A few companies license wind turbine designs to other companies and several companies contract out nacelle manufacturing. Blades and towers may be produced in-house or by outside suppliers. Nacelles, blades, and towers are shipped directly from the manufacturing plant to the construction site.

Nacelles: The plants that produce nacelles are primarily assembly facilities. Nacelle components are produced (in-house or by outside suppliers) to the specifications of the OEM and then assembled at the nacelle plant. The assembly of a nacelle takes less than a week. With high demand in 2008, U.S. plants generally operated two to three shifts per day and five to seven days per week.

Blades: Wind turbine blades are advanced in design, but labor intensive in the manufacturing process. OEMs usually have unique designs for blades. Manufacturing includes labor intensive processes like adding layers of fiberglass to blade molds and finishing the edges of blades. It takes about one week to produce a blade.

Towers: OEMs design the wind turbine towers. Towers usually have 3 sections, each consisting of metal rings that are thickest at the bottom of the tower and are conical in shape since towers taper slightly from the base to a narrower opening at the top. During the manufacturing process plated sheets are cut, rolled into the conical shape and then welded into rings. Rings are then welded together and painted. Platforms, ladders and other accessories are added prior to shipping.

Nacelles, blades, and towers are shipped from the plant directly to the construction site, where they are installed by the project developer. Trucks are the most common method of transport, with specialized trailers required for transporting large pieces like nacelles and blades. Rail and barge shipping are less expensive and avoid the complicated permitting process associated with shipping heavy and oversized products across multiple states, but are used less frequently since they usually do not go to the construction site.

A rotor is installed on a wind turbine (left) and a blade for a GE turbine is transported to the National Renewable Energy Lab (right).

Photos courtesy of DOE/NREL, Credit – Jim Green (left) and Klaus Obel (right).

Sources: Industry officials, interviews by Commission staff, October 29, 2008; industry officials, interviews by Commission staff, October 30, 2008; industry official, interview by Commission staff, Washington, DC, September 24, 2008; industry official, telephone interview by Commission staff, December 10, 2008; "Expanding Inwards," Win[d], May 8, 2008, 4–7; Gamesa Web site. http://www.gamesa.es (accessed January 15, 2008); LM Glasfiber Web site. http://www.lmglasfiber.com (accessed March 5, 2009); DOE, EERE, 20% *Wind Energy by 2030*, 33; Kajrup and Flamholz, "Bending the Wind," August 14, 2003; Port of Houston, 2003 Annual Report, 26-29; Wise, "Port Services are Essential to Wind Energy Logistics: Niche Services are Available in an Emerging Growth Market," March 2006; and Siemens, "Siemens Begins Transportation of Blades by Rail," March 4, 2009.

Many foreign OEMs are localizing nacelle production in the United States in order to take advantage of the growing market, reduce transportation costs,[16] minimize the risks associated with currency fluctuations, ease logistical challenges associated with exporting large nacelles and components, and avoid import duties.[17] Spanish OEMs Acciona Windpower and Gamesa established manufacturing plants in Iowa and Pennsylvania, respectively, and are the only two foreign-based OEMs currently assembling nacelles in the United States.[18] However, six other European OEMs—Emergya Wind Technologies, Fuhrlander, Nordex, Nordic Windpower,[19] Siemens, and Vestas—are planning to open assembly plants.[20]

Box 4. Raw Materials

Steel is the primary material used in wind turbines, but there is also significant use of adhesives, aluminum, blade core materials, concrete, copper, and fiberglass. Rising commodity prices drove up production costs until the fall of 2008, but the financial crisis has put downward pressure on commodity prices.

Wind turbine raw materials, percent by weight, including blades and towers

	Concrete	Steel	Aluminum	Copper	Fiberglass	Adhesive	Core Materials
Weight (%)	1.3	89.1	0.8	1.6	5.8	1.1	0.4

Source: DOE, *20% Wind Energy by 2030*, 63.

Note: Numbers may not add to 100 percent due to rounding.

Supply Chain

The U.S. wind turbine supply chain consists of a combination of in-house production and outsourcing (figure 1), with many European suppliers strongly positioned in the supply chain. Most OEMs manufacturing in the United States want to localize important pieces of the supply chain in order to reduce transportation costs and logistical difficulties, avoid import duties, and mitigate the risks associated with currency fluctuations.[21] As a result, European suppliers are investing in U.S. production and U.S. companies are entering the wind turbine manufacturing industry as suppliers to OEMs.

The rapid growth in global demand in the last few years strained the wind turbine supply chain. In response, some OEMs expanded and diversified their supply chain while others enhanced in-house production capabilities through investments in new manufacturing facilities or purchases of major component suppliers. Different business models have led to different degrees of vertical integration by company and by component. Suzlon, for example, has pursued a strategy of in-house production and vertical integration for most major components. GE is less vertically integrated than Suzlon, leveraging its experience and competitive advantage in supply chain management to build its wind turbine supply chain. Siemens falls in the middle. Unless all production is in-house, companies usually have at least two suppliers for key components.

Nacelle Components

Rising demand in the last few years has led to a shortage of key nacelle components such as gearboxes, large bearings, and large castings (table 2). For example, lead times could be 16 to 18 months for bearings and 40 weeks for gearbox components.[22] There are several components that only a small number of companies are able to produce to the specifications of the wind turbine manufacturing industry and these companies found it difficult to meet sharply rising demand. OEMs' efforts to build their supply chain and/or in-house production capacity, together with the flattening of demand that resulted from the financial crisis, have reduced pressure on the supply chain. Most companies now find that while supplies are tight, they are able to get the components that they need. BTM Consult[23] forecasts that supply chain constraints for key components are likely to abate by 2012.[24]

The supply chain for nacelle components reflects the different sourcing strategies and levels of vertical integration of OEMs. Looking at the same three companies, Suzlon purchased major drive train and gearbox manufacturer Hansen in 2006 and is expanding in-house production of components such as control panels and generators. Siemens purchased Flender and its drive system and generator manufacturing subsidiaries Winergy and Loher, but has not pursued the same extent of vertical integration as Suzlon. GE outsources a larger percentage of components, but does rely on a combination of in- house production and outsourcing for some components (e.g., gearboxes and generators).[25] Other companies demonstrate the same variation in the extent of vertical integration. For example, Acciona does not produce any gearboxes or generators in- house, while Gamesa produces around half of each in-house (figure 2).

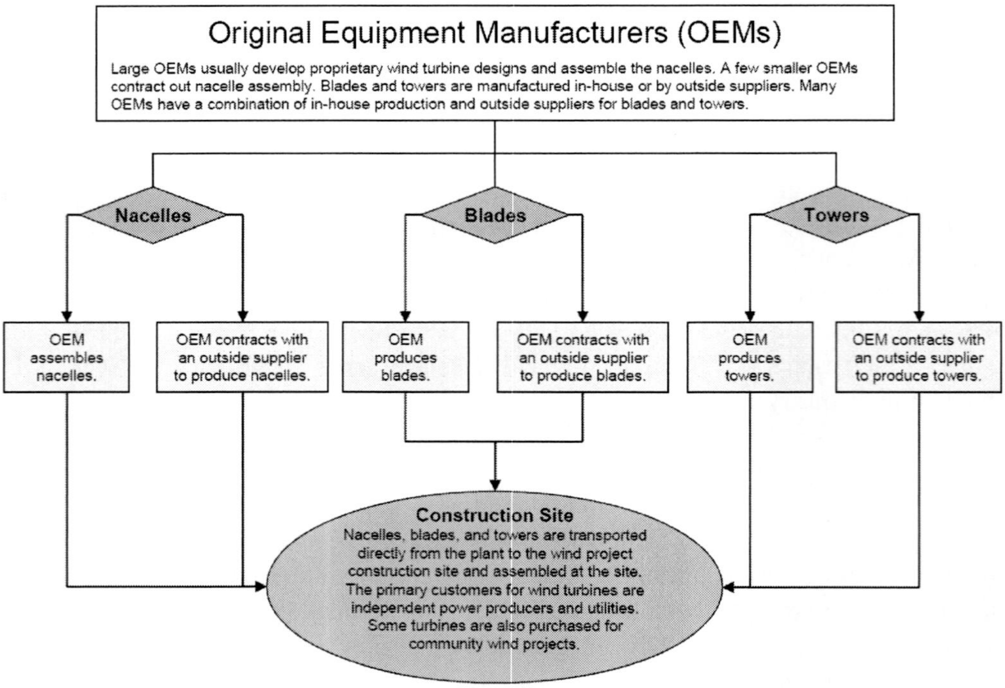

Source: Prepared by USITC staff.

Figure 1. Wind turbine production process.

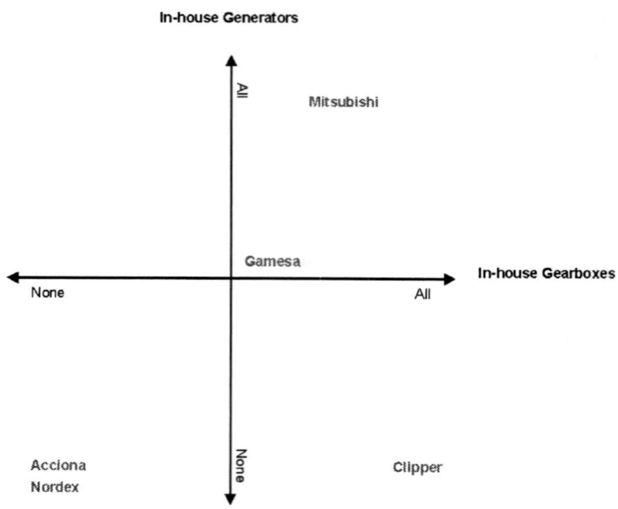

Sources: Compiled from interviews, company documents and Web sites, and Magee, "Wind Turbine Supply Chain."

Notes: Clipper assembles gearboxes on site, but the main components are provided by suppliers. Figures 2–4 are based on companies for which information is publicly available. Not all companies make this information available for all components. Therefore, there is variation in the OEMs used in each example.

Figure 2. Vertical integration, gearbox and generator manufacturing, five companies, 2008.

Table 2. Supply chain status, nacelle components, 2006, 2008, and 2012 forecast

2006		2008	2012 (E)
Bearings	Constraint	Constraint	Exceeds Demand
Castings/Forgings	Constraint	Constraint	Exceeds Demand
Gearboxes	Constraint	Exceeds Demand	Exceeds Demand
Generators	Exceeds Demand	Exceeds Demand	Exceeds Demand

Sources: BTM Consult, "The International Wind Industry 2008–2012," August 11, 2008; EWEA, "Supply Chain: The Race to Meet Demand," November 2006, 27.

Notes: E: Estimated. Constraint is defined here as a shortage or production capacity and demand roughly equal.

The extent of market concentration, shortages, and domestic manufacturing capability varies by component:

Bearings: There is a shortage of bearings for the main shaft, gearbox, and pitch and yaw control. Only a few manufacturers are capable of producing the large bearings needed for wind turbines.[26]

Castings: There is a shortage of casting suppliers in the United States that can make the size needed for the wind turbine manufacturing industry. Currently, most of the suppliers that produce large castings are in Europe, though OEMs are actively looking to qualify local casting suppliers.[27]

Gearboxes: During the peak of wind turbine demand, there was a significant shortage of gearboxes. OEMs indicate supplies remain tight, but that shortages have abated.[28] The international gearbox market is very concentrated, with three European companies, Winergy (now part of Siemens), Hansen (now part of Suzlon), and Moventas, supplying multiple OEMs and controlling a significant share of the market. A group of at least half a dozen smaller, independent companies also supply gearboxes or are entering the market. In order to ensure adequate supplies, some companies have pursued a strategy of vertical integration (though none of the major OEMs meet all of their needs through in- house production). European manufacturers are expanding production in the United States. Moventas plans to build a new plant in Minnesota, Winergy is expanding production in Illinois, and Brevini plans to build a plant in Indiana.[29] Two U.S. manufacturers, K&M Machine Fabricating and Brad Foote, have entered the industry.

Generators: There is limited market concentration among generator suppliers. Generators are usually supplied all or in part by external suppliers and are available in adequate quantities.[30] The leading generator suppliers are European companies such as ABB, Ingeteam, Siemens (via its purchase of Flender), and Elin, though a few OEMs outside of Europe have in-house manufacturing capabilities.[31] In North America there is limited manufacturing capability, with no known investments in generator manufacturing in the United States by European companies.[32]

Sources: Compiled from company documents and Web sites.

Figure 3. Vertical integration, blade manufacturing, four companies, 2008.

Table 3. Representative list of open or planned blade manufacturing plants in the United States, 2009

Company	Location	Headquarters	OEM/Supplier
Gamesa	Pennsylvania	Spain	OEM
Knight and Carver	South Dakota	United States	Supplier
LM Glasfiber	Arkansas, North Dakota	Denmark	Supplier
Molded Fiberglass	South Dakota	United States	Supplier
Nordex	Arkansas	Germany	OEM
Polymarin Composites	Arkansas	Netherlands	OEM
Siemens	Iowa	Germany	OEM
Suzlon	Minnesota	India	OEM
TPI Composites	Iowa	United States	Supplier
Vestas	Colorado (2)	Denmark	OEM

Sources: Complied from AWEA publications, media reports, news releases, and company Web sites.

Notes: This is a representative list and may not include all blade manufacturers. If there is more than one plant in the state, the number of plants is listed in parentheses. Some facilities may also produce other components such as nacelles. Polymarin Composites is a wholly owned subsidiary of Emergya Wind Technologies. Bureau van Dijk, Orbis Companies Database (accessed various dates).

Blades

Many OEMs produce some blades in-house, but the extent of vertical integration varies by manufacturer (figure 3). Some companies outsource all blades (e.g., Clipper), while other OEMs produce all of their blades in-house (e.g., Suzlon). Most companies fall somewhere between these two extremes. OEMs usually have their own blade design and outside suppliers produce blades to their specifications. [33]

The U.S. supply chain for blades includes (1) OEMs with in-house production capabilities, (2) established European suppliers, (3) U.S. companies in related industries that are entering the wind manufacturing industry, and (4) relatively new suppliers outside of Europe and the United States. The first two groups are increasingly meeting U.S. demand by investing in production in the United States and North America. Combined with the entry into the market by U.S. companies in related industries, this has led to substantial growth in domestic manufacturing capabilities.

There are currently at least 12 open or planned blade manufacturing facilities in the United States (table 3). 11 of these 12 plants were opened or announced since 2003. Six OEMs (Gamesa, Nordex, Siemens, Suzlon, Vestas, and Emergya Wind Technologies via subsidiary Polymarin Composites) have or are establishing their own U.S. blade manufacturing facilities. European company LM Glasfiber, the largest global supplier with 25 percent of the world market, has two manufacturing facilities in the United States and one in Canada.[34] At least three U.S. companies, TPI Composites, Molded Fiberglass, and Knight and Carver, have made the transition from related areas to become suppliers to OEMs. TPI Composites and Molded Fiberglass have supply agreements with GE.[35]

While the U.S. manufacturing base has expanded, OEMs also source blades from outside the United States. For example, some OEMs ship blades from Europe, TPI Composites formed a joint venture with Mitsubishi in Mexico that supplies blades for Mitsubishi turbines in the United States, and LM Glasfiber has a manufacturing facility in Canada that can supply blades. In addition, Brazilian company Tecsis is a major supplier to the wind turbine manufacturing industry, providing blades for at least three OEMs in the U.S. market, including blades that it supplies as part of a reported 4 year, $1 billion agreement with GE.[36]

Towers

OEMs tend to be less vertically integrated in tower production than in blade production due, at least in part, to the fact that more companies have the expertise and capabilities necessary to make towers.[37] For this same reason, there are more independent tower producers than blade producers. As with blades, some OEMs outsource all tower production while others have in-house production (figure 4). Companies that produce both blades and towers in-house tend to have a lower percentage of in-house blade production than in-house tower production.

The U.S. supply chain for towers consists of the same four groups that comprise the supply chain for blades, but there is a different allocation of market share between the groups:

(1) *OEMs with in-house production facilities*: OEMs produce towers in-house for the U.S. market, but in-house production makes up a smaller percentage of the U.S. market than in-house production of blades.
(2) *Established European suppliers*: Information on the extent to which European suppliers are exporting to the United States is limited. The proportion of tower manufacturing facilities in the U.S. that are owned by European suppliers is small.
(3) *U.S. manufacturers*: U.S. companies in related industries are entering the market for towers in greater numbers than companies are entering blade manufacturing.[38]
(4) *Suppliers outside of Europe and the United States*: Producers in Canada, Mexico, and Asia have significant U.S. market share. At least one Asian company has invested in U.S. production.

U.S. tower production has expanded rapidly, primarily due to investment by OEMs and U.S. companies in related industries. There are at least 15 companies producing or planning to produce towers at 20 locations (table 4). Two European OEMs and one European tower manufacturer have established or are planning to open U.S. manufacturing plants. Eleven U.S. companies manufacture towers at 16 U.S. plants and one U.S. company established a joint venture with a Japanese company to manufacture towers in the United States.

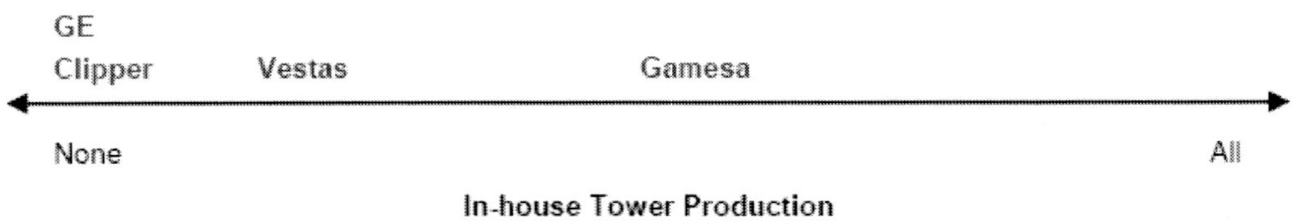

Sources: Compiled from Magee, "Wind Turbine Supply Chain," December 9, 2008 and company documents and Web sites.

Figure 4. Vertical integration, tower manufacturing, four companies, 2008.

Table 4. Representative list of open or planned tower manufacturing plants in the United States, 2009

Company	State	Headquarters	OEM/Supplier
Aerisyn Inc	Tennessee	United States	Supplier
Ameron	California	United States	Supplier
Bergen Southwest Steel	Texas	United States	Supplier
DMI	North Dakota, Oklahoma	United States	Supplier
Dragon Wind	Colorado	United States	Supplier
Gamesa	Pennsylvania	Spain	OEM
Katana Summit	Nebraska	Japan and United States (joint venture)	Supplier
Martifer	Texas	Portugal	Supplier
Northstar Wind Towers	Nebraska	United States	Supplier
RTLC Windtower	Texas	United States	Supplier
SMI & Hydraulics	Minnesota	United States	Supplier
T. Bailey	Washington	United States	Supplier
TowerTech (Broadwind)	South Dakota, Texas, Wisconsin	United States	Supplier
Trinity Structural Towers	Illinois, Iowa, Texas	United States	Supplier
Vestas Wind Systems	Colorado	Denmark	OEM

Sources: AWEA publications, news releases, media reports, and company Web sites.
Note: This is a representative list and may not include all tower manufacturers.

OEMs often source towers locally due to high transportation costs and the risk of exchange rate fluctuations, but OEMs also source towers for the U.S. market from Canada, Mexico, and Asia. Towers from Asia may be cheaper for projects on the West Coast than towers made at distant U.S. plants and, by maintaining suppliers in Asia, OEMs have more flexibility in their supply chain.[39] China, Indonesia, Korea, and Vietnam are among the major producers of wind towers for the U.S. market. For example, Dongkuk S&C in Korea and CS

Wind Tower Company, a Korean company with manufacturing in Vietnam and China, have supplied Vestas and the U.S. market.[40] In Canada, there are at least three companies (Hitachi, DMI Industries, and Marmen) capable of supplying towers for the U.S. market.[41] Trinity Structural Towers is capable of supplying the U.S. market from Mexico.[42]

Entering the Supply Chain

There is substantial interest among U.S. manufacturers outside the wind turbine manufacturing industry in becoming suppliers to OEMs, a trend that has accelerated as companies negatively affected by the financial crisis and the recession look to diversify their customer base. But the financial crisis has made it more difficult to enter the industry. While OEMs want to add local suppliers, becoming a supplier is expensive and challenging due to the need to reinvest and retool plants and to demonstrate the ability to expand as the OEMs grow. The current financial environment makes it difficult to access the capital necessary to overcome these barriers.[43] The American Recovery and Reinvestment Act (Stimulus Bill) attempts to make it easier for manufacturers to enter wind turbine and other renewable manufacturing industries by creating a 30 percent tax credit for retooling manufacturing facilities or investing in new manufacturing plants.[44]

OEMs note that potential suppliers are often unfamiliar with the large size of wind turbine components and the reliability and quality requirements.[45] Wind turbine components must meet a high quality threshold given the twenty year life expectancy of the turbine and expense of replacing components in an installed turbine. As a result, there is an extensive qualification process that can take 2 to 3 months for simple parts and 12 to 15 months for more complicated parts. Some companies have representatives on-site at suppliers around the world to ensure that products meet quality requirements.[46]

Despite these challenges, most OEMs are confident in the skills and capabilities of U.S. industry and the ability of U.S. companies to make the transition to wind.[47] Anecdotal evidence and data for individual states indicate the growth in nacelle, blade, and tower manufacturing has led to growth in the number of companies in the U.S. producing materials or components for nacelles, blades, and towers. For example, Clipper Windpower produced its first wind turbines in Iowa in 2006 and now has about 90 instate suppliers. The number of wind turbine manufacturing industry suppliers in Ohio increased from fewer than 12 in 2004 to about 37 in 2007 to more than 50 in 2008.[48] In Colorado, Hexcel and Creative Foam Corp. are establishing manufacturing plants to supply the new Vestas blade plants.[49]

Location of Nacelle, Blade, and Tower Plants

Proximity to markets and access to multiple shipping options are two of the most important factors for manufacturers in deciding where to establish manufacturing facilities. Due to the size and weight of nacelles, blades, and towers, proximity to markets is important in reducing shipping costs, minimizing logistical difficulties, and avoiding permitting challenges that increase with the distance from the construction site.[50] The largest markets in the United States (figure 5) tend to be in the central United States in close proximity to the best and most widely available onshore wind resources (figure 6) or on the West Coast. Therefore, most manufacturing plants are located near large markets (figure 5).

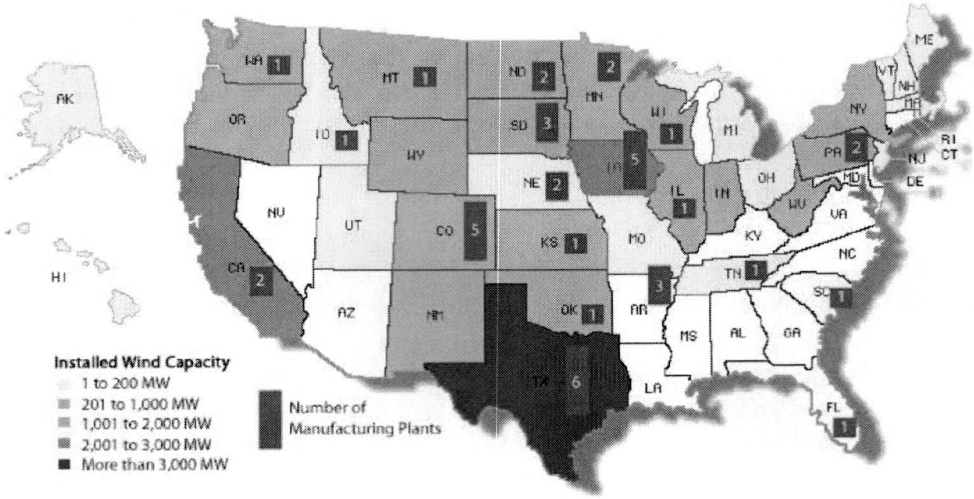

Sources: AWEA Web site. http://www.awea.org/projects (accessed various dates); see Tables 2, 3, and 4 for manufacturing plant data; and map template from DIY Maps Web site. http://monarch.tamu.edu/~maps2/ (accessed various dates).

Note: In several cases, blades and nacelles are produced in the same plant. These plants are only listed once. Based on a representative list of manufacturing plants. The map may not include every plant in the United States. However, it does provide a good representation of the distribution of manufacturing activities.

Figure 5 Installed wind capacity and location of blade, nacelle, and tower manufacturing plants, by state, March 2009.

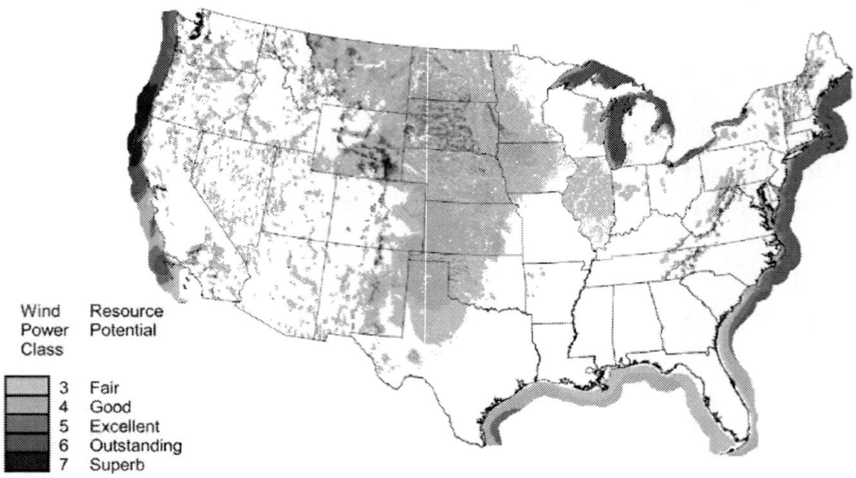

Source: Department of Energy, Wind Powering America Web site. http://www.windpoweringamerica.gov (accessed various dates).

Notes: Areas with marginal (class 2) and poor (class 1) wind resources are not shown. While some of the best wind resources are off the East and West Coasts, no offshore turbines have been installed in the United States.

Figure 6. National Renewable Energy Lab (NREL) wind resources map.

Table 5. Reported wind turbine manufacturing job losses, November 2008–March 2009

	Plants w/Layoffs	Plants Closed	States with Layoffs	Original Employment at Affected Plants	Approximate Layoffs
Nacelle Plants	2	0	Iowa (2)	578	148
Blade Plants	3	1	Arkansas (2), Pennsylvania, South Dakota	1440	360
Tower Plants	3	1	North Dakota, Oklahoma (2), Tennessee	779	295
Total	**8**	**2**		**2,797**	**803**

Sources: Compiled from press releases and media reports.

Note: Number in parentheses indicates the number of plants in the state affected (if more than one). Layoffs at one of the nacelle plants include global layoffs for that OEM. However, most of these layoffs were expected to affect the U.S. manufacturing plant.

Companies also consider the availability and adequacy of shipping options when making location decisions. Companies ship most often by truck and want to be close to a good interstate system, but multiple shipping options are also important. Therefore, companies generally locate near rail lines and/or waterways that offer lower cost shipping options. Proximity to suppliers and other companies in the wind turbine manufacturing industry are also considerations.[51]

The quality of the workforce, availability and quality of worker training programs, proximity to good community colleges, traditional employment base (e.g., history of manufacturing), and labor and operating costs are also important in location decisions.[52] Political support, positive relationships with state and local officials, and financial incentives are also considerations, and state and local governments have offered extensive incentives to attract these industries. State and local support includes property tax abatements, sales tax reductions, low interest loans, and support for worker training. In some cases, local governments have made investments in infrastructure, such as extending railroads and improving highway ramps. Some states have offered support for R&D and the deployment of new technology.[53]

Employment

Employment in wind turbine manufacturing increased from 2003 to 2008, but was negatively affected by the financial crisis and recession in late 2008 and early 2009. In 2008, manufacturing employment was about 20,000, a significant increase from employment in 2007.[54] The number of wind related jobs among GE suppliers increased from about 2,000 in 2008 to over 4,000 in 2009.[55]

The recession and the financial crisis led to layoffs in late 2008 and early 2009. At least eight wind turbine manufacturing industry plants laid off workers and two plants closed between November 2008 and March 2009 (table 5). As a result, 803 out of 2,797 jobs (29 percent) at these plants were lost.[56] In addition to the job losses, planned expansions of several plants were postponed and a few companies delayed new plant construction. However, not all jobs losses were due to the recession and the financial crisis. Gamesa laid off about 180 workers involved in blade production at one Pennsylvania plant with the intention of expanding production at another U.S. plant that can make larger blades. LM Glasfiber laid off about 150 employees and closed one of its two plants in Arkansas, but this was a facility that was originally intended to be temporary and was only kept open due to the high level of demand in 2008.[57] In addition, many planned investments remain unaffected and are still moving forward. For example, Vestas broke ground on nacelle and blade factories in Colorado in March 2009.[58] In total, new manufacturing facilities announced in 2008 could lead to about 4,000 additional jobs in the long term.[59]

Manufacturers (nacelle, blade, and tower) generally seek skilled workers in areas such as mechanical and electrical assembly or workers with specific skills that translate into wind energy manufacturing (e.g., previous manufacturing or related experience).[60] Manufacturers often look for locations with community college systems that provide good technical training for workers. In addition, manufacturers train workers on site and multinational corporations often have exchange programs with employees from overseas coming to newly opened U.S. plants and U.S. employees going overseas for training.[61]

Based on 12 manufacturing plants for which information was publicly available, average hourly wages in manufacturing plants (nacelle, blade, and tower) are generally between $13 and $20 (table 6).[62] The median of the average hourly wage rate at the twelve companies was $15.

Research and Development

As an emerging technology that is on the verge of price competitiveness with traditional sources of power, research and development (R&D) is critical to the wind turbine industry. OEMs are developing and testing (a) new multi-MW onshore wind turbines, (b) offshore wind turbine models, and (c) turbines that can operate in low wind or very cold climate conditions. Through their R&D programs, OEMs also seek to (1) optimize nacelle, blade, and component designs and materials, (2) improve the reliability of wind turbines, (3) improve turbine technology, and (4) tailor turbines for local markets. Corporate spending on R&D varies and comparisons across companies may be unreliable given potentially different definitions and accounting methods; however, reported expenditures do give a sense for spending on R&D. Vestas spent $175 million (119 million Euros, 2.0 percent of revenue) [63] on R&D in 2008, Suzlon spent $3.4 million (0.2 percent of sales) on R&D in the fiscal year ending in 2008, and Clipper spent $21.1 million on R&D in 2008 (3 percent of revenue). U.S. companies GE and Clipper have R&D capabilities in the United States and Siemens and Vestas plan to open U.S. R&D facilities.[64]

Table 6. Estimated average hourly wages at 12 manufacturing plants

	Sample Size	Average Hourly Wage Range
Nacelle Plants	3	$15 to $20
Blade Plants	6	$13 to $17
Tower Plants	2	$15 to $16
Nacelle and Blade Plants	1	$17
Total	**12**	**$13 to $20**

Sources: Compiled from press releases and media reports and Kelly, "Wind Energy Update," June 26, 2008.

Note: The date of the reported wages varies, but they are generally from 2007 and 2008.

Government programs and public-private collaboration through national laboratory systems aid the development of wind technology and the testing of new wind turbine models. The United States generally budgets more money for wind energy research, development, and demonstration (RD&D) than other countries with competitive wind industries (figure 7). From 2003 to 2007, the average annual U.S. wind energy RD&D budget was $44 million, while Germany's was $19 million, Denmark's was $16 million, and Japan's was $9 million.[65] However, U.S. funding has been highly inconsistent over time and is currently lower than funding for solar and biomass research, which received an average of $102 million and $114 million, respectively, from 2003 to 2007. Comparing renewable energy RD&D expenditures as a whole to nuclear and fossil fuel RD&D expenditures, cumulative renewable energy RD&D expenditures from 1974 to 2007 were significantly less than RD&D expenditures for nuclear energy and fossil fuels and, on an annual basis, were lower throughout most of the period. In 2007, the U.S. RD&D budget for nuclear energy was $629 million, for fossil fuels $367 million, and for renewable energy $416 million.[66]

Globalization

OEMs have entered global markets through a combination of exports and overseas production. OEMs often prefer to establish overseas production facilities since there are high transportation costs and logistical challenges associated with exporting nacelles, blades, and towers.[67] According to a recent survey of turbine manufacturers, more than 40 percent of companies indicated that the best way to enter new markets is through joint ventures. Another 40 percent indicated that the best way is through a subsidiary, while less than ten percent prefer to export.[68] As a reflection of this, large OEMs and blade manufacturers tend to produce in at least two of the major global markets, though tower manufacturers are less likely to produce outside their home region (figure 8).

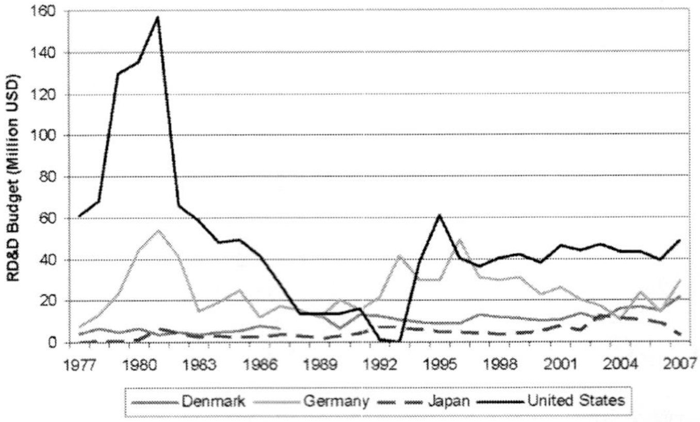

Sources: International Energy Agency (IEA), IEA Energy Technology R&D Statistics Service (accessed various dates).

Notes: 1988 data for Denmark are not available. Data are in constant 2007 U.S. dollars at the 2007 exchange rate.

Figure 7. Wind energy RD&D budget for select countries, 1977–2007.

	Asia		Europe					North America		
	China	India	Denmark	Germany	Poland	Portugal	Spain	Canada	Mexico	U.S.
Acciona	Nacelles						Nacelles (2) Blades			Nacelles
Clipper										Nacelles
Gamesa	Nacelles Blades						Nacelles (4) Blades (5) Towers (4)			Nacelles Blades Towers
GE	Nacelles			Nacelles			Nacelles	Nacelles		Nacelles (3)
Siemens			Nacelles Blades (2)							Nacelles Blades
LM Glasfiber	Blades (2)	Blades (2)	Blades (2)		Blades		Blades (3)	Blades		Blades (2)
Molded Fiberglass										Blades
TPI Composites	Blades								Blades (2)	Blades
DMI								Towers		Towers (2)
Martifer						Towers				Towers
Trinity Industries									Towers	Towers (3)

Source: Compiled from company Web sites and media reports.

Notes: Nacelles highlighted in gray, blades in blue, and towers in beige. The number in parentheses represents the number of manufacturing plants in the country (if there are multiple plants). Several of the plants are under construction. This figure is limited to companies that manufacture in the United States. As a result, it may appear to over represent the percent of global production that takes place in the United States. It does not include all companies that manufacture in the United States.

Figure 8. Global production locations, eleven companies, 2009.

Despite the preference of OEMs for local production, it is not feasible to establish production facilities in every market and large increases in global trade in the last few years indicate that companies can profitably export wind turbines and components. Companies make decisions about whether to export or establish local production facilities based on factors such as the proximity of the market to existing plants, transportation costs, comparative labor and production costs, the local supply base, the favorability and stability of the exchange rate, trade barriers and duties, and the size and stability of the market.[69]

Among U.S. producers, two strategies are emerging. First, U.S. manufacturers, including both U.S.-based companies and foreign companies with U.S. manufacturing facilities, are beginning to export products, primarily to smaller, emerging markets in North and South America. Many of the new U.S. manufacturing plants were established with the intention of serving both the U.S. market and other markets in North and South America.[70]

Second, U.S. companies are investing in foreign production facilities in key wind energy markets. GE has manufacturing plants (some inherited from its purchase of Enron Wind) in Canada, China, Germany, Spain, and the United States.[71] The facilities are located in large wind markets and, in the case of Canada, Spain, and China, countries with local content requirements at the local or national level.[72] Clipper established an R&D center in the United Kingdom to design a 7.5 MW offshore wind turbine that may be also produced in the United Kingdom.[73]

Some U.S.-based companies view the Chinese market as providing the best opportunity outside the United States. Several U.S.-based producers have established manufacturing locations in China in order to enter the market. For example, GE established a nacelle plant in China; blade manufacturer TPI Composites established a Chinese manufacturing plant following an agreement to supply blades to GE; and Timken established a joint venture in China to produce bearings.[74] Despite this growth, wind turbine production in China by U.S. companies is significantly less than production by European companies.[75]

U.S. Market

The United States wind turbine market is rapidly expanding and is now the largest in the world in terms of annual wind turbine installations. Annual installed capacity increased from 1,672 MW in 2003 to 8,545 MW in 2008 (figure 9).[76] Wind energy makes up a growing portion of new electricity generation capacity, accounting for 35 percent of capacity additions in 2007.[77] Wind energy accounts for a small but growing portion of total U.S. electricity generating capacity, increasing from less than one percent in 2003 to 1.7 percent in 2007.[78] The Department of Energy has set a goal of generating 20 percent of U.S. electricity from wind by 2030.[79]

Currently, the U.S. market is almost entirely for onshore wind turbines since they are less expensive to install and maintain, and easier to site than offshore wind turbines. In the United States, there are also ample available onshore sites. However, offshore wind turbine installations in the United States are likely to grow in the long term since the United States has excellent offshore wind resources close to major population centers.[80]

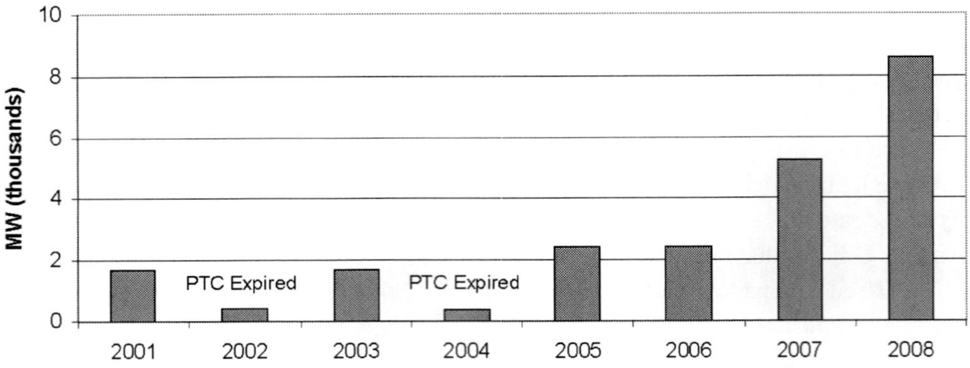

Source: AWEA, *Annual Wind Industry Report*, 10.

Note: Wind turbine installations declined in 2002 and 2004 because the production tax credit (PTC), one of the primary incentives for wind installations, expired.

Figure 9. Annual installed wind energy capacity, 2001–08.

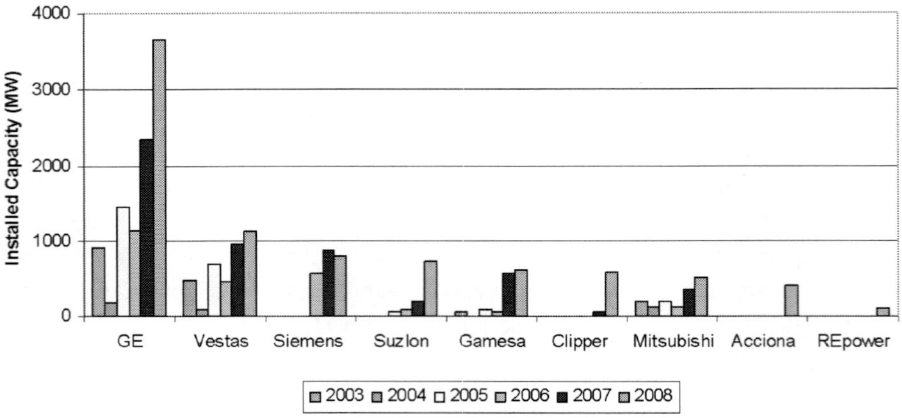

Source: AWEA.

Notes: Does not include companies with 10 MW or less installed. Vestas installations in 2003 include turbines manufactured by NEG Micon, which subsequently merged with Vestas.

Figure 10. Installed wind turbines by manufacturer, 2003–08.

Wind Turbine Suppliers and Market Competition

The number of OEMs in the U.S. market is expanding as the size of the domestic market grows. The number of OEMs in the U.S. market increased from five in 2003 to 13 in 2008.[81] U.S.-based GE maintained its position as the leading OEM in the U.S. market from 2003 to 2008, but market growth and the market entry of additional OEMs reshaped the market share of competitors. In 2003, the three leading competitors in the U.S. market were GE, Vestas, and Mitsubishi, with 55, 30, and 12 percent of the market, respectively (figure 10). Each of these companies had strong sales growth between 2003 and 2008, but the market share of each company also decreased due to other companies entering the market and companies already in the market increasing their market share. GE installations increased from 903 MW in 2003 to 3,657 MW in 2008 (305 percent), though its market share decreased from 55 percent to 43 percent. Vestas was second in the U.S. market in 2008 with 1,120 MW installed, followed by Siemens (791 MW), and Suzlon (736 MW). Gamesa was fifth in the market in 2008 with 616 MW, followed by U.S.- based Clipper Windpower (595 MW), which had strong sales growth in only its second year of sales. Mitsubishi fell from third to seventh in installed capacity from 2003 to 2008, despite installations increasing from 201 to 516 MW. New market entrants Acciona (410 MW) and REpower (102 MW) had strong sales. Four other companies (Fuhrlander, DeWind, AWE, and Baoding Huide) entered the U.S. market for the first time and had installations of 10 MW or less. Nordex is opening a U.S. plant and was active in the U.S. market in prior years.[82]

Several other OEMs have indicated their intention to sell turbines in the United States, suggesting that the number of competitors in the wind turbine market will further increase. Nordic Windpower and Emergya Wind Technologies have announced plans for domestic manufacturing facilities that can serve the U.S. market. AAER, a Canadian company, recently started production in Quebec and signed several agreements in 2008 to supply individual 1.5 MW turbines to U.S. customers. Hyundai licensed designs from a subsidiary of American Semiconductor Corporation (AMSC) for a 1.65 MW turbine that it plans to start producing for the U.S. market.[83] Hitachi, which manufactures wind turbine towers in Canada, plans to introduce a 2.0 MW turbine, designed with Fuji Heavy Industries, into the U.S. market in the next few years.[84] Vermont based Northern Power Systems is developing a 2.2 MW turbine for the U.S. market.[85] Even if some of these companies do not successfully enter the U.S. market, it seems likely that the number of competitors will increase in the next few years.

As new companies enter the market and OEMs invest in U.S. production, competition in the U.S. market will likely continue to increase. Most OEMs maintain proprietary designs, and wind turbines are differentiated by a number of factors including price, reliability, capacity, efficiency, and availability. Three of these factors, capacity, reliability, and price, are discussed below.

Capacity: The wind turbine market is shifting toward larger wind turbines capable of generating more electricity. The average size of wind turbines installed in the United States increased from 0.71 MW in 1998–99 to 1.65 MW in 2007 and the number of turbines over 1 MW increased from 0.3 percent of wind turbines in 1998–99 to 89 percent in 2007.[86] GE's 1.5 MW turbine is the most commonly installed turbine. The largest turbine installed in 2008 was 3.0 MW.[87]

Reliability: With an expected life of 20 to 25 years, wind turbine reliability is an important concern for customers. Wind turbine maintenance is expensive, both in terms of the cost of repairs and lost generating time, so customers want to buy turbines that will require minimal maintenance.[88]

Price: U.S. wind turbine prices went through a substantial period of decline in the 1980s and 1990s, reaching a low point of about $700/kW ($700,000/MW) between 2000 and 2002, according to an analysis by Lawrence Berkeley National Lab. Since that time, turbine prices have increased to as much as $1 ,240/kW in 2007 ($1,240,000/MW) due to rising prices for commodities (e.g., copper and steel), tight supply of key components, and a weak U.S. dollar. In addition, rising turbine demand and tight supply gave OEMs more pricing power.[89] Wind turbine prices appear to have leveled off in 2008 and may decrease due to the effects of the financial crisis (the decline in demand and material prices), improvements in the supply chain, and expansions in global manufacturing capacity. Vestas, for example, significantly expanded its manufacturing capacity to meet rising global demand, but now has excess capacity. Project developers are indicating that it is easier to secure turbines than it was before the credit crisis and that they expect OEMs to have less pricing power in the next few years.[90]

Wind Turbine Customers

The primary customers for wind turbines are independent power producers (IPPs) and utilities, with some community wind farms also purchasing turbines (box 5). These customers purchase wind turbines primarily for commercial electricity generation and, in many cases, place large orders, as evidenced by the fact that the average size of wind farms completed in 2008 was about 70 MW.[91] At the end of 2007, IPPs owned 84 percent of cumulative installed wind capacity, utilities owned 14 percent, and community ownership was 2 percent. The four largest owners[92] of wind farms at the end of 2008 were U.S.-based IPP NextEra Energy Resources (formerly FPL Energy) with 6,290 MW, Spain-based IPP Iberdrola Renewables with 2,063 MW, U.S.-based MidAmerican Energy with 1,940 MW, and Portugal-based IPP Energias de Portugal with 1,873 MW.[93]

Box 5. Wind Project Terminology

Community Wind Farms: Wind projects that have community ownership rather than ownership by a utility or an independent power producer.

Independent Power Producers (IPPs): A company that produces power that it sells to electric utilities.

Institutional Tax Equity Investors: Firms that invest in wind energy to utilize the PTC to offset other sources of income. Examples include GE Financial Services, New York Life, Morgan Stanley, and Wells Fargo.

Power Purchase Agreement: A long-term agreement for the sale of energy from a power producer to a utility.

Wholesale Power Prices: The price of electricity sold by electric generators to utilities.

Note: For more on institutional tax equity investors, see Wiser and Bolinger, *Annual Report*, 14.

Wind turbine customers also include large power users, like universities and military bases, that install turbines to generate electricity for their own use. These customers often purchase a single turbine with a power output suited to their energy demand. For example, the U.S. Marine Corps Logistics Base in Barstow, California installed a single AAER wind turbine.[94] For some companies, such as Nordic Windpower and AAER, marketing products to these customers is an important part of their business strategy.

Factors Affecting Wind Turbine Demand

The primary wind turbine customers, utilities and IPPs, have a choice of power generation options. IPPs and utilities are investing in wind as one of the options to meet a portion of rising domestic electricity demand. Wind is an abundant, clean energy source, with wind turbine installations closely tied to government mandates for renewable energy, the ability to finance wind projects, and the cost-competitiveness of wind energy. The ability to finance projects and the cost-competitiveness of wind energy are closely tied to federal tax policies. These factors affect wind turbine installations and, therefore, the demand for wind turbines.

Government Mandates

One significant factor that has spurred the development of wind energy is the passage of renewable portfolio standards (RPS) at the state level. RPS mandate a certain percentage of electricity from renewable sources (usually 10 to 25 percent) by a particular date (generally between 2015 and 2025). Twenty-eight states and the District of Columbia have passed RPS (figure 11) and five states have established voluntary goals.[95] These RPS encourage the construction of wind farms by IPPs and utilities, thereby increasing demand for wind turbines.

In addition, utilities are investing in wind and other renewable energy sources as a way to hedge against the possibility of future national carbon regulation. One study found that utilities in the West are considering the potential impact of carbon regulation and, as a result, are including investment in renewable energy in their plans.[96]

Electricity Prices

The federal production tax credit (PTC), the maturation of wind technology, and rising fossil fuel prices have made wind energy increasingly competitive with prices from traditional energy sources. On a national basis, wind power prices are competitive with wholesale power prices. Wind power prices decreased from $63/MWh in 1999 to $40/MWh in 2007, with prices reaching a low point in 2005 and rising slightly in 2006 and 2007 due to rising wind turbine prices and higher costs for newly installed wind projects.[97] Once a wind project is installed, the price of power is usually stable over time, while changes in fossil fuel prices can affect the price of electricity from coal and natural gas plants. Compared to the annual range of wholesale wind power prices from 2003 to 2007, wind power prices were either below or at the low end of the range each year.[98] One of the key factors in reducing wind energy prices to the point where they are competitive with wholesale power prices is the inclusion of the PTC, a tax credit for the generation of renewable energy that is equivalent to about 2.1 cents/kWh over the twenty year life of a wind project.[99]

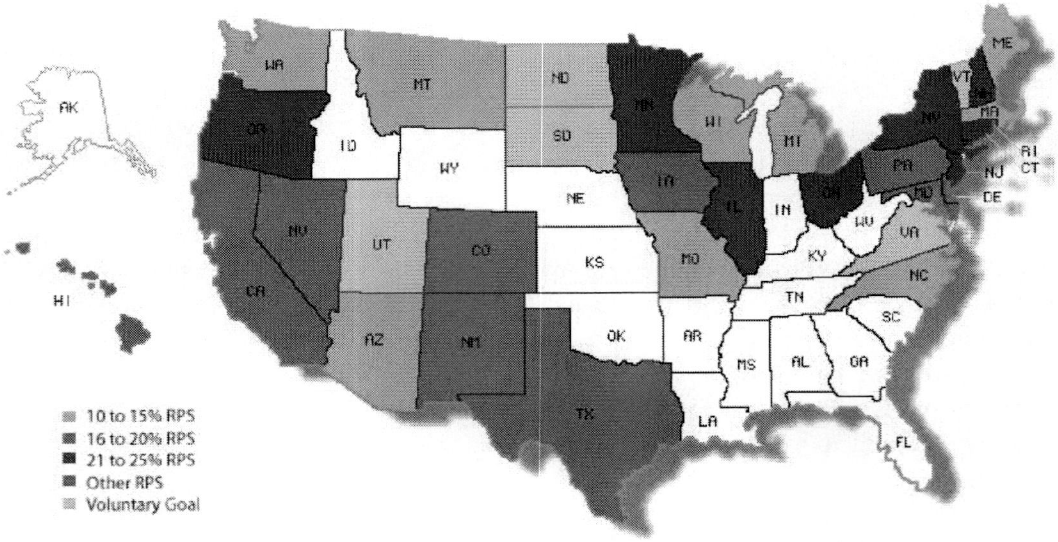

Source: North Carolina Solar Center, NC State University. Database of State Incentives for Renewables and Efficiency (accessed various dates); map template from DIY Maps Web site. http://monarch.tamu.edu/~maps2/ (accessed various dates).

Note: Iowa's RPS is 105 MW and Texas's RPS is 5,880 MW by 2015. Colorado, New Mexico, and Oregon have lower standards for certain small, cooperative, and/or municipally owned utilities. Massachusetts requires 15 percent of electricity from renewable sources by 2020. The Massachusetts RPS is scheduled to increase by one percent per year after 2020. The District of Columbia's RPS is 20 percent by 2020

Figure 11. State renewable portfolio standards.

There may be variations in the competitiveness of wind energy in different regions since wholesale power prices are not the same everywhere in the country. At the regional level, the cost competitiveness of wind energy will vary depending on factors such as the primary type of electricity generation (e.g., coal, natural gas, nuclear) currently in use, quality of wind resources, and costs of installing wind turbines.[100] The price competitiveness of wind energy for project developers has a significant impact on wind turbine installations and, consequently, domestic wind turbine demand.

Financing

Project developers need access to significant capital in order to finance wind projects. There are a number of different methods that are used to finance large wind projects in the United States. Some projects are financed through loans or bonds. Increasingly common in the lead up to the financial crisis, however, were a variety of financing mechanisms that involved investment by institutional tax equity investors. Institutional tax equity investors are firms that invest in wind power projects in order to use the PTC to offset other income. These tax investors are a significant source of financing for wind energy projects.[101]

The financial crisis demonstrates the impact of financing constraints on wind project construction in the United States. As noted earlier, the financial crisis has led to a decrease in wind project construction and a consequent downturn in demand for wind turbines. The financial crisis has made it more difficult for project developers that finance projects themselves to finance new wind energy projects. However, projects that rely on investment by institutional tax equity investors who use the PTC to offset income and reduce taxes are the most severely impacted. The declining financial health of many of these investors has resulted in a rapid decline in the use of the PTC and a shrinking in the number of tax equity investors. With less money available, the pace of development will likely slow in the short term.[102]

This has led to a decrease in demand for wind turbines, at least in the short term. The pace of new orders has slowed and some OEMs have reported that developers have delayed or cancelled orders.[103] In the longer term, OEMs expect demand to revive as developers seek to meet state RPS, tax equity investment begins to return, and companies start to take advantage of the incentives in the Stimulus Bill.[104] In addition, the negative effects of the financial crisis in the U.S. market will likely be somewhat offset by declining prices for turbines and increased supply of turbines.[105] The Stimulus Bill has a number of provisions that are directly targeted at mitigating the impact of the financial crisis on wind project financing:

Production Tax Credit (PTC): Extends the expiration date of the PTC from the end of 2009 to the end of 2012.

Investment Tax Credit (ITC): Makes wind energy project developers eligible for either a 30 percent tax credit for investment in new wind energy projects or a federal grant equal to the value of the ITC.

Accelerated Depreciation: Extends accelerated depreciation for wind projects through 2009.

Other: Provides for other loan guarantees, bonds, and R&D funding.[106]

Other

Several other factors, including transmission capacity, the intermittence of wind energy, policy stability, transportation permitting, and wind farm siting, affect wind farm construction and, therefore, demand for wind turbines. Many of these factors may be impediments to the long-term growth of the wind industry.

Transmission Capacity: Because the major wind corridor is in the central United States, wind projects are often distant from the areas with the highest electricity demand. There is currently insufficient transmission capacity to bring this wind energy to population centers. There are also issues regarding access to the grid for renewable energy and in transmitting energy across regional grids. Smart grid technologies[107] that can better handle the variable loads from renewable energy generation are also needed. While there is clear consensus on the need for new transmission capacity, there are difficulties in siting new lines and uncertainty over who pays for them.[108] The Stimulus Bill includes some funding for transmission projects and grid modernization.[109]

Intermittence: Wind is an intermittent energy source, with the amount of energy generated varying depending on the available wind resources at a given time. While some studies indicate that this intermittence can be reasonably handled and more wind energy can be integrated into the electric grid, wind would benefit from a viable energy storage system.[110]

Policy Stability: Government policy has played an important role in spurring wind installations, but the production tax credit has expired several times since it was first passed in 1992. Uncertainty over the future of the PTC inhibits long-term planning and decreases the willingness of companies to invest in wind energy, which has a negative impact on the supply chain and domestic manufacturing capability.[111] An industry study illustrates the importance of the PTC for the wind energy industry—in the three years when the production tax credit lapsed (2000, 2002, and 2004), wind turbine installations fell by 73 to 93 percent (figure 9).[112] In 2008, the PTC was extended through the end of 2009. The Stimulus Bill extends the PTC through the end of 2012, providing for a period of continuity in wind energy policy.

Transportation and Permitting: A complicated and decentralized permitting process makes it difficult to arrange for the transportation of major turbine components to construction sites. Companies must apply for a permit in each state through which they transport nacelles, blades, and towers, and the ease of the application process varies by state. In addition, states and localities have different requirements, such as what time of day and days of the week products can be transported, and how many blades can be shipped per truck, that complicate transportation logistics.[113]

Siting: There is generally broad public support for the development of wind energy. A recent survey found that 82 percent of Americans would support a wind project in their community.[114] Landowners (often farmers and ranchers) are generally supportive of wind projects because they are compensated for use of their land and can receive roughly $3,000 to $6,000 per MW per year.[115] However, local residents are often concerned about the impact of a wind project on the local landscape, property values, wildlife, and the economy.[116]

U.S. Trade

The United States is a net importer of wind turbines and major components. U.S. imports increased significantly from 2003 to 2008, while exports rose only slightly. The U.S. trade deficit for wind-powered generating sets (box 6) increased from $355 million in 2003 to $2.5 billion in 2008. The United States has the largest trade deficits with Denmark, Spain, Japan, Germany, and India, which are home to the major global turbine manufacturers.[117]

U.S. Imports

Wind-Powered Generating Sets (HTS 8502.31.0000)

Imports of wind-powered generating sets increased from $356 million in 2003 to $2.5 billion in 2008 (more than 600 percent) as the major European and Asian wind turbine producers entered or expanded their presence in the growing U.S. market (table 7). Five

European countries (Denmark, Spain, Germany, the United Kingdom, and Portugal) and two Asian countries (Japan and India) accounted for 98 percent of all U.S. imports in 2008.[118] Denmark and Spain accounted for a combined 55 percent of imports. Denmark continued to be the largest single source of imports, but imports from Denmark decreased from $966 million in 2007 to $707 million in 2008 while imports from Spain increased from $423 million to $669 million. Japan was the third largest source with 15 percent of imports ($378 million) in 2008, and Germany was fourth with 12 percent ($297 million).[119]

Box 6. HTS Classification of Wind Turbines

The discussion of trade data in this report focuses on imports and exports of wind-powered generating sets. Wind turbines are classified under "wind-powered generating sets" (8502.31.0000) in the Harmonized Tariff Schedule of the United States (HTS). This provision includes wind turbine nacelles and components, such as blades and hubs, when they are imported with the nacelle. When imported separately, blades, hubs, generators, and other components are classified under other HTS provisions. This HTS provision includes both small and utility scale wind turbines, but small wind turbines generally represent a low percentage of wind turbine trade.

Imports of components will be discussed for three groups of goods—(1) generators (HTS provision 8501.64.0020), (2) towers (7308.20.0000), and (3) blades and other components (8412.90.9080 and 8503.00.9545). These components are classified under provisions that also include goods that are not used in wind turbines. But wind accounts for a significant portion of trade in each category and appears to be a major driver of import growth. A similar analysis of component exports is not included since wind exports represent a much smaller percentage of trade in these categories.

Table 7. U.S. imports of wind-powered generating sets (HTS 8502.31.0000), 2003–08, thousand dollars

	2003	2004	2005	2006	2007	2008	Percent Change 2003–08
Denmark	215,045	31,708	328,169	630,818	966,111	706,864	229
Spain	38,552	702	44,758	157,519	423,492	669,079	1,636
Japan	85,297	24,337	62,569	97,228	307,475	377,836	343
Germany	301	18	166	56,543	210,660	297,212	98,486
India	15,636	0	12,482	216,572	253,487	178,598	1 042
United Kingdom	57	2,903	23,390	39,689	129,789	137,110	240,765
Portugal	0	0	0	0	0	80,446	
Italy	3	0	8,667	651	0	33,721	1,148,817
China	40	14	36	7,997	85,248	14,327	35,830
All Other Countries	709	278	2,252	1,648	3,676	8,157	1,049
Total	**355,641**	**59,960**	**482,489**	**1,208,667**	**2,379,940**	**2,503,349**	**604**

Source: Compiled from official statistics of the U.S. Department of Commerce.

Wind-powered generating set imports are correlated with wind turbine installations (figure 12). From 1998 to 2005, changes in wind-powered generating set imports were closely correlated with changes in wind turbine installations. That correlation appears to have subsequently weakened and, despite a significant increase in wind turbine installations in 2008, imports only slightly increased. Among the possible explanations are that many of the turbines installed in 2008 were imported in 2007, or that a rise in domestic manufacturing led to a reduction in imports. Rising domestic production is suggested by an increase in imports of AC generators—which are used in domestic wind turbine nacelle assembly (see below).

Generators (HTS 8501.64.0020)

The rise in domestic production of nacelles has led to an increase in imports of generators for those nacelles. Imports of AC generators[120] with outputs from 750 to 10,000 kVA (the size most common in utility scale turbines) increased from $69.6 million in 2003 to $142.8 million in 2006 and then to $644.5 million in 2008 (figure 13).[121] Major import sources in 2008 were Denmark, Germany and Japan, which are home to major OEMs and generator manufacturers. Spain was the second largest source of imports in 2007, but U.S. imports from Spain declined by almost 90 percent from 2007 to 2008.[122]

Blades & Other Components (HTS 8412.90.9080 and 8503.00.9545)

U.S. imports of parts of motors and generators (which include blades and other turbine components)[123] increased from $363 million to $1.8 billion from 2003 to 2008 (figure 14).[124] There are significant imports of these components from European countries (Denmark, Germany, and Spain) that are home to major OEMs, blade manufacturers, and component producers. Imports from these three suppliers increased 565 percent from $87 million in 2003 to $576 million in 2008.[125] Brazil, Canada, and Mexico also are major sources of U.S. imports, likely because they are home to manufacturing plants for large independent blade producers, such as Tecsis (Brazil), the TPI-Mitsubishi joint venture VienTek (Mexico), and LM Glasfiber (Canada).[126] Imports from Brazil, Canada, and Mexico increased almost 400 percent from $154 million in 2003 to $763 million in 2007.[127]

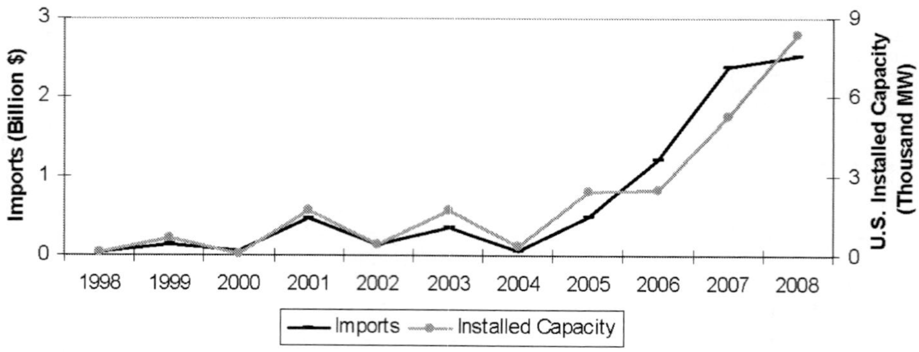

Sources: Compiled from official statistics of the U.S. Department of Commerce; AWEA, *Annual Wind Industry Report*, 4.

Figure 12. U.S. wind-powered generating set imports and installations, 1998–08.

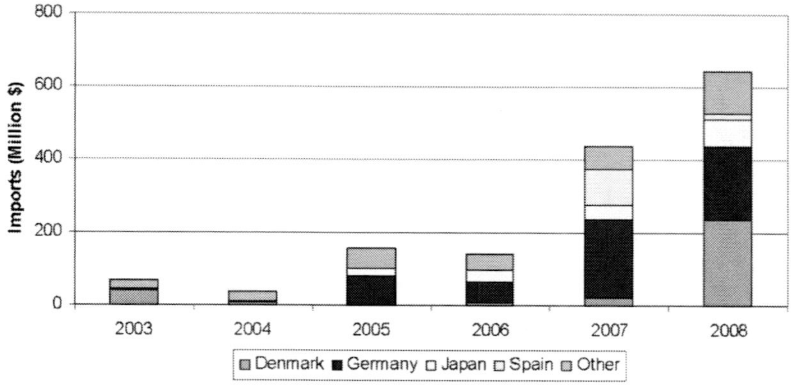

Source: Compiled from official statistics of the U.S. Department of Commerce.

Figure 13. U.S. imports of AC generators, 750–10,000 kVA (HIS 8501 .64.0020), 2003–08.

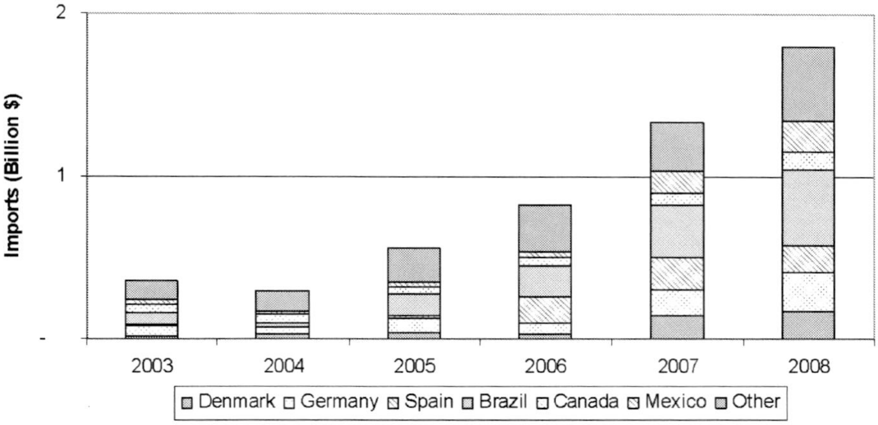

Source: Compiled from official statistics of the U.S. Department of Commerce.

Figure 14. U.S. imports of parts of motors and generators (HTS 8412.90.9080 and 8503.00.9545), 2003–08.

Towers (7308.20.0000)

Towers are one of the components for which manufacturing capabilities are most readily available and the supply chain is the most diverse, with significant U.S. imports from Europe (Denmark and Spain), Asia (China, Indonesia, Korea, and Vietnam), Canada, and Mexico (figure 15).[128] Imports of towers and lattice masts increased from $41 million in 2003 to $944 million in 2008.[129] Asia, Canada, and Mexico are increasingly important in the supply chain for towers. Four Asian countries, China, Indonesia, Korea, and Vietnam, account for a combined 62 percent of towers and lattice masts imports in 2008. Canada and Mexico accounted for 23 percent of imports.[130] The share of imports from Denmark and Spain fell from 34 percent in 2007 to 13 percent in 2008, primarily due to a drop in imports from Denmark. Imports from Denmark increased from $2 million to $128 million in 2007 before declining to $44 million in 2008, possibly due to a change in sourcing strategy by OEMs.[131]

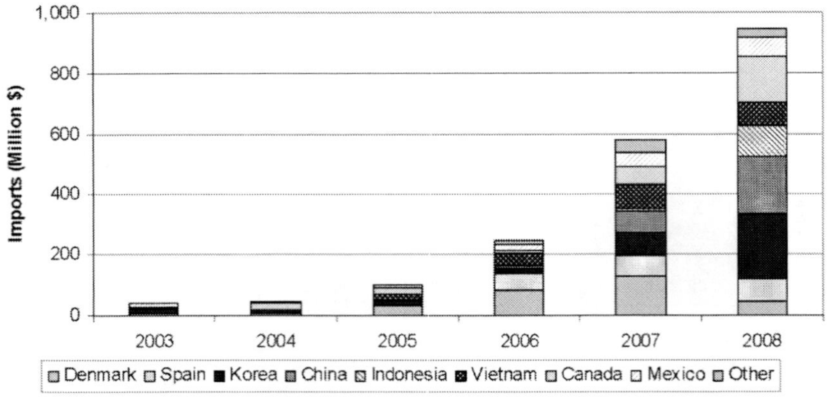

Source: Compiled from official statistics of the U.S. Department of Commerce.

Figure 15. U.S. imports of towers and lattice masts (7308.20.0000), 2003–08.

Table 8 U.S. exports of wind-powered generating sets (HTS 8502.31.0000), 2003–08, thousand dollars

	2003	2004	2005	2006	2007	2008
China	0	0	0	0	12,837	16,202
Brazil	0	0	0	0	0	5,871
St Kitts-Nevis	0	0	0	0	675	0
Japan	742	0	0	0	646	0
Canada	0	0	0	83,310	0	0
Germany	0	4,398	3,626	0	0	0
All Other Countries	4	0	0	0	0	0
Total	**746**	**4,398**	**3,626**	**83,310**	**14,158**	**22,073**

Source: Compiled from official statistics of the U.S. Department of Commerce.

U.S. Exports

U.S. exports of wind turbines were limited and sporadic, but anecdotal evidence suggests that export and overseas investment opportunities are growing. Exports of wind-powered generating sets increased from $746,000 in 2003 to $22.1 million in 2008 (table 8). Exports to China increased from $0 in 2006 to $16.2 million in 2008 and exports to Brazil rose from $0 in 2007 to $5.9 million in 2008. The only exports to Canada between 2003 and 2008 were in 2006, when exports totaled $83.3 million.[132]

There is anecdotal evidence that exports of turbines and blades are slowly expanding, particularly within the Americas. For example, Acciona Energy plans to source turbines for at least one of the wind farms that it is developing in Canada from its Iowa factory.[133] Clipper signed an agreement in 2008 to supply 27 wind turbines for a project in Mexico and CTC/DeWind exported ten 2 MW wind turbines to Chile in December 2008.[134] Nordic Windpower has an order for three 1 MW turbines for a wind farm in Uruguay.[135] In 2008, 48 blades were shipped from a North Dakota factory to Australia.[136] LM Glasfiber is reportedly shipping blades produced at its Arkansas facility to South America, and tower and other blade manufacturers have established U.S. manufacturing plants with the possibility of supplying markets throughout the Americas.[137]

U.S. and Foreign Trade Measures

U.S. Tariff and Nontariff Measures

The general rate of duty for wind turbines and components ranges between free and 3 percent (table 9). The rate of duty for wind-powered generating sets is 2.5 percent; for AC generators 2.4 percent; and for other parts of generators 3 percent. Towers and other parts of engines and motors (including some blades) enter free of duty.[138] There are no known U.S. nontariff measures (NTMs) that affect U.S. imports of wind turbines and components.

U.S. Government Trade Related Investigations

One patent infringement investigation related to wind turbines occurred during the period from 2003 to 2008 and one earlier order remains in effect. GE filed a complaint in February 2008 alleging violations of section 337 of the Tariff Act of 1930. The complaint alleges that Mitsubishi Heavy Industries, Ltd. and two related or subsidiary companies violated this act by importing variable speed wind turbines and components that infringe on GE patents. This case is pending.[139] An earlier complaint by Kenetech Windpower against Enercon GmbH resulted in an exclusion order prohibiting imports of certain unlicensed Enercon variable speed wind turbines and components until the expiration of the patent on February 1, 2011.[140] Enercon and GE, the current owner of the patent, reached a cross-licensing agreement in October 2004, but the exclusion order for the United States remains in effect.[141] As of December 2008, there are no orders in place or ongoing investigations as a result of countervailing duty or anti-dumping cases.

Table 9. Wind-turbines and components: Harmonized Tariff Schedule subheading, description and column 1 rate of duty as of January 31, 2009

HTS Subheading	Article Description	Rates of Duty
		General
7308.20.0000	Towers and lattice masts	Free
8412.90.9080	Other parts of engines and motors	Free
8501.64.0020	AC generators of an output exceeding 750 kVA but not exceeding 10,000 kVA	2.4%
8502.31.0000	Wind-powered electric generating sets	2.5%
8503.00.9545	Other parts of generators for use with machines of heading 8501 or 8502	3.0%

Source: Harmonized Tariff Schedule of the United States.

Foreign Tariff Measures

There is significant variation in foreign tariff rates for wind-powered generating sets (table 10). Duties on wind powered-generating sets range from free (e.g., Australia, Canada, Japan, Mexico, and South Africa) to 10 percent or more (e.g., India and Taiwan). The European Union's 2.7 percent tariff rate is similar to the 2.5 percent tariff rate applied by the United States. According to a recent World Bank study, applied tariffs on wind-powered generating sets, wind turbine generators, and towers average 3 percent among high-income WTO members (table 11). Among low- and middle-income countries, the average applied tariffs on generators and wind-powered generating sets are 5 percent. Duties on imported towers are somewhat higher at 10 percent.[142] Several countries have adjusted duties to encourage the import of components rather than fully assembled nacelles.[143]

Table 10. Applied tariff rates, wind-powered generating sets, 2008 unless otherwise noted

Country	Tariff Rate	Country	Tariff Rate
Australia[a]	Free	Japan	Free
Canada	Free	Mexico	Free
China[b]	8%	South Africa	Free
European Union	2.7%	South Korea[c]	8%
India	16%	Taiwan	10%

[a] Generating sets exceeding 500 kVA.
[b] As of 2004.
[c] As of 2004.
Sources: Compiled from International Customs Tariffs Bureau, APEC Tariff Database, and schedules of individual countries.

Table 11. Wind turbine bound and applied tariff rates, by income level, percent

HS Code	Article Description	Low- and Middle-Income WTO Members		High-Income WTO Members	
		Maximum Average Bound Tariffs	*Average Applied Tariff Rates*	*Maximum Average Bound Tariffs*	*Average Applied Tariff Rates*
7308.20	Towers and lattice masts	28	10	16	3
8501.64	AC generators of an output exceeding 750 kVA but not exceeding 10,000 kVA	28	5	16	3
8502.31	Wind-powered electric generating sets	26	5	16	3

Source: *World Bank, International Trade and Climate Change: Economic, Legal, and Institutional Perspectives*, 2008, 130–1 32.

Foreign Nontariff Measures

Local content requirements implemented at the national level in some countries and at the regional level in other countries are important nontariff measures.[144] Local content requirements generally mandate that a certain percentage of the value of installed wind turbines be produced locally. These requirements encourage the development of a local wind turbine manufacturing industry by making local producers more competitive and/or encouraging foreign producers to locate production in the market rather than exporting products to the market. Several of the countries with local content requirements (Canada, China, and Spain) will be discussed in more detail below.

FOREIGN INDUSTRY PROFILES

European OEMs have historically dominated the global wind turbine manufacturing industry and, along with GE, lead the industry today. Three of the four largest suppliers of wind turbines in 2008 were based in Europe (figure 16). Vestas (Denmark), Gamesa (Spain), and Enercon (Germany) ranked first, third, and fourth, respectively, in global wind turbine installations in 2008. U.S.-based GE, which also has manufacturing and assembly facilities in Canada, China, Germany, and Spain, and wind research centers around the world, was the second largest supplier of wind turbines. These four companies were the four largest global suppliers in 2004 as well. During 2004 to 2008, however, GE gained global market share due to the growth of the U.S. market and became the world's second largest supplier.[145] Three other European companies, Siemens, Acciona, and Nordex, are also among the top ten suppliers of wind turbines.[146]

Another significant change from 2004 to 2008 was the growth of Asian turbine manufacturers. Chinese companies Sinovel and Goldwind are now among the top ten suppliers of wind turbines due to robust demand within China. Indian manufacturer Suzlon gained market share and is the fifth largest manufacturer. Chinese OEM Dongfang and Japanese OEM Mitsubishi are the eleventh and thirteenth largest manufacturers, respectively.[147]

Trade statistics confirm that European companies are strongly positioned in the wind turbine manufacturing industry, but that exports from other countries are also increasing. Denmark's exports of wind-powered generating sets increased by 29 percent, from $966 million in 2003 to $1.2 billion in 2008, but its global export share fell from 80 percent to 23 percent (figure 17). Germany's exports increased by over 1,100 percent from $164 million to $2.0 billion and its export share increased from 14 to 38 percent. Exports from India (12 percent of exports in 2008), Japan (9 percent), Spain (9 percent), and China (4 percent) also significantly increased.[148]

Government policy often plays an important role in the development of a successful industry. According to Lewis and Wiser, policies that support the development of local manufacturing are often important in the development of a successful wind industry. Policies that governments have implemented to support turbine manufacturers include (1) policies that limit competition from products manufactured outside the country (e.g., local content requirements and import duties) and (2) programs to encourage the development of industry (e.g., funding for research and development, tax incentives, and subsidies for the purchase of locally made turbines).[149] In addition, a large, stable home market is often important in the early stages of industry development and governments usually play an important role in the creation of this market through policies such as feed- in tariffs,[150] renewable portfolio standards, tax credits, loans and subsidies, and utility or government power purchase agreements.[151]

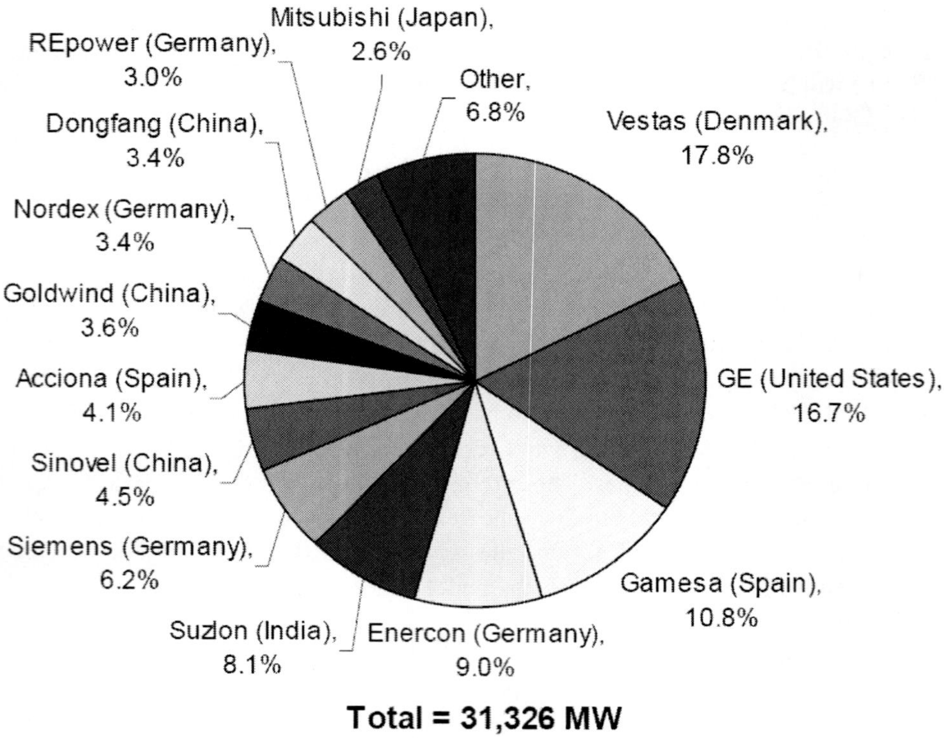

Sources: BTM Consult, *International Wind Energy Development: World Market Update 2008*, 24.

Note: BTM Consult data are based on the number of turbines delivered by OEMs. This may differ from the total number of turbines installed because not all turbines delivered in a particular year are installed in the same year. BTM Consult generally calculates market share as the number of turbines delivered in a year by a particular company divided by total global wind turbine installations. Since the number of installations often differs from the number of turbines delivered, this can result in a total market share for companies that is more or less than 100 percent. For the purposes of this report, which is concerned with manufacturing, the total number of turbines delivered, as reported by BTM Consult, is divided by the total global deliveries of wind turbines to calculate global market share. See BTM Consult, *International Wind Energy Development: World Market Update 2008*, 30.

Figure 16. Leading suppliers of wind turbines, 2008.

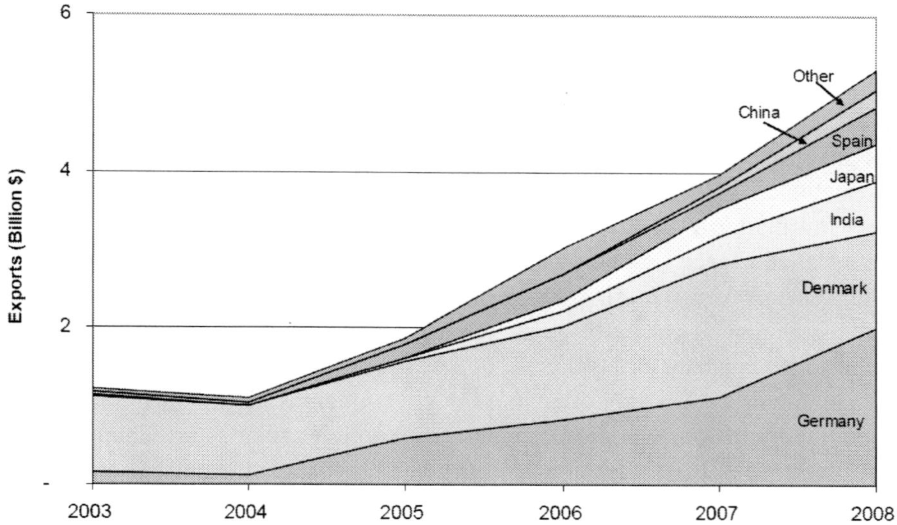

Source: GTIS, Global Trade Atlas (accessed various dates).

Figure 17. Exports of wind-powered generating sets by country, 2003–08.

Table 12. Global exports and employment of the wind industries in Denmark, Germany, and Spain

	Exports					Direct Employment
	2003		*2008*		*Percent Change*	
	Exports (USD)	*% of Global Exports*	**Exports Exports (USD)**	*% of Global Exports*		
Denmark	965,937,124	79.5	1,243,393,519	23.4	29	23,500
Germany	164,019,083	13.5	2,004,311,102	37.7	1,122	38,000
Spain	45,811,192	3.8	469,680,570	8.8	925	17,769

Source: Global Trade Atlas (accessed March 8, 2009); EWEA, *Wind at Work*, 7; Spanish Wind Energy Association (AEE), *Wind Power 2008, Sector's Yearbook: Analysis and Data*, 57.

Notes: Export data are exports of wind-powered generating sets. Direct employment typically includes a range of activities such as manufacturing, operations and maintenance, and R&D, but there may be some variation in the definition of direct employment for the employment figures above.

The importance of these factors[152] is illustrated in the following profiles of the wind turbine manufacturing industries in three of the countries which are the most significant producers of wind turbines, Denmark, Germany, and Spain (table 12).[153] The profiles focus on OEMs, but each of these countries is also a major source of components.

Denmark

Wind turbine manufacturers in Denmark benefited from both active government support and their early entrance into the wind market. Danish manufacturers began entering the wind turbine industry in the 1970s, often transitioning from other industries and benefiting from the availability of workers in declining industries, including agriculture and shipbuilding. Danish companies initially entered the growing wind energy market in California, but this market collapsed in the 1980s and companies came to depend more on their home market. Denmark's government supported the development of the domestic market through a variety of policies including a feed-in tariff and subsidies for the installation of wind turbines.[154] Denmark's wind market was one of the largest in the world and Denmark is still eighth in cumulative installed capacity, but only 77 MW was added in 2008 due to the saturation of the onshore market. However, there is potential for future market growth due to strong offshore wind potential. Denmark currently generates 20 percent of its electricity from wind and cumulative installed capacity at the end of 2008 was 3,180 MW.[155]

Local manufacturing was supported through government R&D funding, and locally made turbines were favored for use in wind farms through policies such as providing financing for wind farms that used domestic turbines and developing stringent standards and certification requirements that benefited local producers.[156] Denmark developed several large companies such as Vestas, NEG Micon, and Bonus, but through mergers and acquisitions the number and the ownership of companies has changed. Vestas and NEG Micon merged and now operate under the Vestas name and Bonus was acquired by Siemens in 2004. Indian manufacturer Suzlon established its international headquarters in Denmark in part to take advantage of available skilled labor.[157] Blade manufacturer LM Glasfiber is headquartered in Denmark. Direct employment in the Danish wind industry was about 23,500 in 2007.[158]

Danish manufacturer Vestas had 17.8 percent of the global market in 2008. German manufacturer Siemens, which maintained a significant manufacturing presence in Denmark following its acquisition of Bonus, held 6.2 percent of the global market.[159] LM Glasfiber is the largest global independent producer of blades, with 25 percent of the global market.[160] Total turnover[161] of the wind industry in Denmark increased from $3.0 billion in 2003 to $7.8 billion in 2007.[162] Denmark's exports of wind-powered generating sets increased from $966 million in 2003 to $1.2 billion in 2008.[163] Denmark's exports peaked at $1.7 billion in 2007.[164]

Germany

The German wind turbine manufacturing industry developed as a result of a combination of policies that supported the development of wind turbine technology and created a market for wind turbine manufacturers.[165] Germany's wind turbine manufacturing industry is one of the largest in the world with major producers such as Enercon, Fuhrlander, Nordex, REpower, and Siemens.[166] Foreign OEMs, including GE and Vestas, also have manufacturing plants in Germany.

With slowing German demand and an increase in the size of the global market, German exports are rising.[167] In 2007, total value added for German manufacturers (including OEMs and suppliers) was $8.4 billion ($6.1 billion Euros), 83 percent of which was exported.[168] Germany's exports of wind-powered generating sets increased from $164 million in 2003 to $2 billion in 2008.[169] Enercon had 9 percent of the global market in 2008, Siemens 6.2

percent, Nordex 3.4 percent, and REpower 3.0 percent.[170] Total direct employment in the wind energy sector in Germany is 38,000.[171]

Spain

Spain has a substantial wind energy manufacturing industry, including several OEMs with varying degrees of domestic and international presence. These OEMs benefit from a feed-in tariff and provincial government regulations requiring local manufacturing.[172] The largest OEM is Gamesa, which had 10.8 percent of the global market in 2008.[173] Gamesa was originally formed as a joint venture with Vestas, but split from Vestas due to strategic differences and the desire to sell turbines outside of the domestic market. Following its split with Vestas, Gamesa acquired Spanish competitor Made.[174] The other major Spanish global supplier is Acciona, which had 4.1 percent of the global market in 2008.[175] The third largest manufacturer is Ecotecnia, which had 0.8 percent of the global market.[176] There are a number of other domestic companies with small shares of the Spanish wind energy market.

The top two global manufacturers of wind turbines (Vestas and GE) have manufacturing facilities in Spain, as does the leading independent manufacturer of blades (LM Glasfiber). In all, Spain has at least 18 factories that manufacture or assemble nacelles and at least a dozen factories producing blades.[177] Direct employment in the wind energy industry in Spain in 2007 was about 17,769, including more than 5,000 manufacturing jobs. Indirect employment, including component suppliers, was an estimated 27,187.[178]

Spain's domestic market has grown,[179] but manufacturers are also becoming increasingly export oriented.[180] Spain's exports of wind-powered generating sets increased from $45.8 million in 2003 to $469.7 million in 2008.[181] Spanish companies are actively investing in overseas production, particularly in growing markets such as China and the United States. Acciona established nacelle assembly plants in China and the United States, and Gamesa has invested in the production of blades, nacelles, and towers in the United States and blades and nacelles in China.[182]

Foreign Market Profiles

Europe continues to lead the world in annual global installed capacity, but a growing percentage of installations are in Asia and North America. The five largest markets in 2008 were the United States, China, India, Germany, and Spain (table 13). These five countries were also the five largest markets in 2007, though Spain fell from second tofifth place. Primarily as a result of the growth of the market in China and India, Asia and the Pacific's market share increased from 25 percent in 2006 to 34 percent in 2008 (figure 18). The North American market share increased from 21 to 33 percent. Europe's market share decreased from 51 percent to 33 percent. Estimates by the Global Wind Energy Council (GWEC) show stability or growth in annual wind turbine installations in all regions, but indicate a further shift in market share toward the Asia-Pacific. The GWEC estimates that Asia will account for 47 percent of all wind turbine installations by 2012, compared to 26 percent for North America and 23 percent for Europe.[183]

The global market for wind turbines has rapidly increased and global imports of wind-powered generating sets increased from $1.4 billion in 2003 to $6.6 billion in 2008.[184] The United States is now the largest market for wind turbines and accounted for a significant percentage of the global increase in imports of wind-powered generating sets from 2003 to 2008 (figure 19).[185]

Table 13. Top 10 countries in annual installed capacity in 2008, MW

		2007	2008			2007	2008
1.	United States	5,244	8,358	6.	Italy	603	1,010
2.	China	3,449	6,300	7.	France	888	950
3.	India	1,730	1,800	8.	UK	427	836
4.	Germany	1,667	1,665	9.	Portugal	434	712
5.	Spain	3,522	1,609	10.	Canada	386	526

Source: GWEC, *Global Wind 2008 Report*, 10, 13.

Note: GWEC data on annual installations may vary slightly from other estimates.

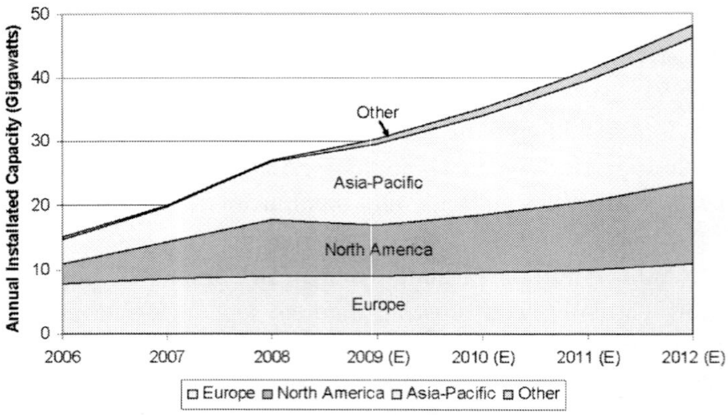

Source: GWEC, *Global Wind 2006 Report, 9; GWEC, Global Wind 2007 Report, 8; and GWEC, Global Wind 2008 Report*, 13.

Note: E: Estimated.

Figure 18. Annual wind turbine installations, 2006–08, and GWEC forecasts through 2012.

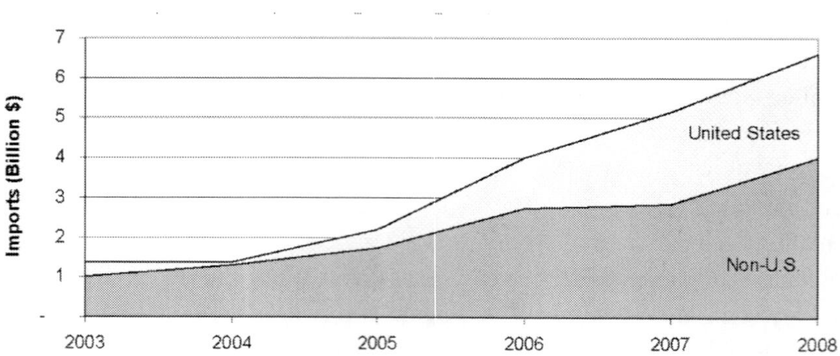

Source: GTIS, Global Trade Atlas (accessed various dates).

Figure 19. Global imports of wind-powered generating sets, 2003–08.

Though the United States continues to be the largest importer of wind-powered generating sets, import growth in other countries was strong. Non-U.S. imports of wind-powered generating sets increased by almost 300 percent from 2003 to 2008. The largest non-U.S. importers in 2008 were Germany, Canada, the United Kingdom, Turkey, and Spain (table 14).[186] Of the 15 largest importers, eight were located in Europe, four in the Asia-Pacific, and three in the Americas. With the exception of Germany, imports by each of the top fifteen importers increased from 2003 to 2008.[187]

Four of the five countries with the most wind turbine installations in 2008, the United States, China, Germany, and Spain, were also among the top ten importers of wind- powered generating sets in 2008. India ranked third in installations, but was 34th in wind turbine imports.[188] A recent survey of wind turbine manufacturers by the Germany Wind Energy Institute asked companies to rank the most important current and future markets for wind turbines. Manufacturers ranked Germany as the most important current wind turbine market and the second most important future market and China as the fourth most important current market, but the most important future market. Spain ranked sixth and eighth, respectively.[189] The next section will profile the market for wind turbines and related components in China, Germany, and Spain as well as in Canada and emerging markets in the Americas.[190]

Table 14. Top 15 importers of wind-powered generating sets, U.S. dollars, 2003–08

		Imports *2003*	*2008*	**Percent Change 2003–08**	**Rank: 2008 Wind Turbine Installations**
1	United States	355,641,053	2,585,636,227	627	1
2	Germany	632,539,838	563,464,674	-11	4
3	Canada	29,694,267	539,654,533	1,717	10
4	United Kingdom	86,586	426,017,776	491,917	8
5	Turkey	731,751	283,650,236	38,663	14
6	Spain	11,430,742	275,967,022	2,314	5
7	Australia	10,300,625	216,691,118	2,004	12
8	Italy	4,042,710	191,069,500	4,626	6
9	China	31,297,102	189,281,571	505	2
10	Japan	98,477,229	174,826,254	78	13
11	France	500,677	139,371,658	27,737	7
12	Portugal	40,310,019	130,001,882	223	9
13	Netherlands	67,897,794	126,861,737	87	11
14	Brazil	1,996,946	121,720,898	5,995	22
15	South Korea	6,485,870	102,169,320	1,475	26
	All others	91,297,229	529,604,925	480	
	Total	1,382,730,438	6,595,989,331	377	

Source: GTIS, Global Trade Atlas (accessed various dates); *GWEC, Global Wind 2008 Report*, 13; EWEA, "Wind Now Leads EU Power Sector."

Note: U.S. import data may vary slightly from official Department of Commerce statistics.

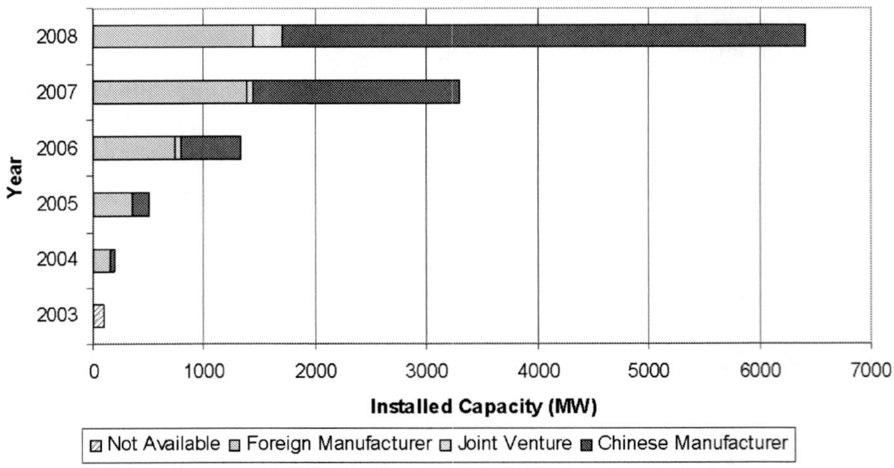

Sources: BTM Consult, *International Wind Energy Development: World Market Update 2008,* 91; GWEC, Global Wind 2007 Report, 51.

Figure 20. Annual installed wind capacity in China, by type of producer, 2003–08.

China

China is one of the fastest growing wind markets in the world, with annual wind energy installations increasing by more than 6,000 percent, from 98 MW in 2003 to 6,300 MW in 2008 (figure 20).[191] The market for wind energy has grown due to China's need to meet its vast and expanding energy needs and the role of the government in supporting the development of the wind market. All wind projects of 50 MW or more are auctioned to project developers by the National Development and Reform Commission, while smaller projects are approved by provincial governments.[192] The government has set goals for wind installations, which have been adjusted several times as previous goals were exceeded. The latest goal is to install 10 GW of wind capacity by 2010 (a goal that was exceeded in 2008).[193] The government has also established a requirement that power producers generate 3 percent of the electricity from renewable sources (excluding hydroelectric) by 2010 and 8 percent by 2020.[194] While the Chinese government has helped spur the development of wind energy, challenges remain, including the need to modernize the electric grid, build transmission lines, and address policies related to the price of electricity.[195]

The Chinese government has implemented a number of policies to encourage the development of local manufacturing to supply the wind energy market. The government has imposed local content requirements, requiring that turbines use 70 percent locally made products, and adjusted import duties to encourage the import of components rather than finished nacelles.[196] The government also encouraged local manufacturing through support for R&D and policies to promote technology transfers.[197]

Competition in the Chinese market has intensified as local companies enter the wind turbine manufacturing industry and major international OEMs enter the market. Foreign companies (e.g., GE, Gamesa, Suzlon, and Vestas) are entering the market primarily through the establishment of subsidiaries and manufacturing facilities within China. Other firms established joint ventures (e.g., Acciona, Nordex, and REpower) and a few firms licensed technology to Chinese companies.[198]

There are at least 40 Chinese OEMs, which have a growing share of the domestic market. Their market share increased from 25 percent in 2004 to 74 percent in 2008.[199] The leading OEMs in the market in 2008 were Chinese companies Sinovel, Goldwind, and Dongfang with 22, 18, and 16 percent of the market, respectively. Due to their large domestic market share, Sinovel, Goldwind, and Dongfang were seventh, ninth, and eleventh, respectively, in global market share. The combined domestic market share of other Chinese manufacturers was 18 percent. Among foreign manufacturers, Vestas was the leading supplier in 2008 with 9 percent of the Chinese market followed by Gamesa with 8 percent, GE and Suzlon with 3 percent, and Nordex and Acciona with 2 percent.[200] Global suppliers, such as blade suppliers LM Glasfiber and TPI Composites, have established manufacturing facilities in China to supply OEMs.[201]

Germany

Government policy in Germany in the 1990s spurred the creation of one of the largest global wind energy markets. In 1991 the German government implemented a feed-in tariff that specified a minimum price for renewable energy and guaranteed access to the grid for renewable energy generation. The tariff applies for 5 to 20 years, depending on the quality of the wind at the site, and declines by a small percentage each year. Combined with regulations to ease siting, national and EU renewable energy goals, and other policies, the feed-in tariff led to the rapid growth of the wind energy market in Germany.[202] Cumulative installed capacity grew from 334 MW in 1993 to 23,895 MW in 2008, and wind energy accounted for 6.4 percent of electricity consumption in 2007.[203]

Germany continues to have one of the largest global wind energy markets, but the market has slowed and is shifting to larger onshore and offshore turbines. Annual installations in 2008 were 1,665 MW, significant but lower than the peak of over 3,000 MW in 2002.[204] Despite the decrease in annual installations, the German market is likely to remain strong due to feed-in tariffs, renewable energy goals, and significant potential wind resources. There are opportunities for expanding onshore installations, developing offshore wind, and repowering existing wind turbines, though neither of the last two accounted for a major portion of installed capacity in 2007.[205]

The German market is dominated by Enercon and Vestas, which accounted for a combined 65 percent of all wind turbines installed in Germany from 1982 through 2008 and a combined 83 percent of all turbines installed in 2008 (figure 21). GE has a manufacturing presence in Germany and a declining, though traditionally strong, market share in Germany. GE accounted for 9 percent of wind turbines installed in Germany from 1982 to 2008, but less than 1 percent in 2008.[206] That is likely to change since GE recently completed upgrading its plant in Germany to produce GE's new 2.5 MW turbine and already has at least one order in Germany.[207]

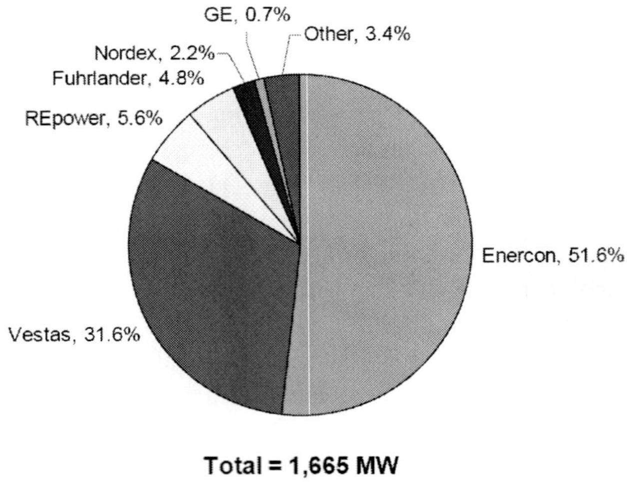

Source: Ender, "Wind Energy Use in Germany–Status 31.12.2008," February 2009, 54.
Note: Numbers may not add to 100 percent due to rounding.

Figure 21. Wind turbine installations in Germany, by OEM, 2008.

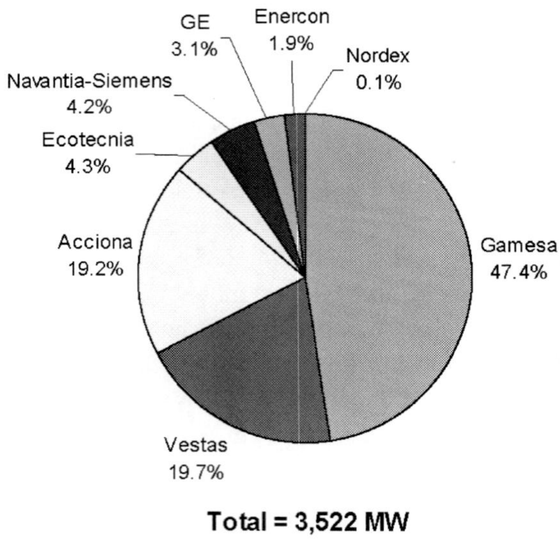

Source: AEE, *Wind Power 2008*, 16-17.
Note: Numbers may not add to 100 percent due to rounding.

Figure 22. Wind turbine installations in Spain, by OEM, 2007.

Spain

Spain established a feed-in tariff for wind energy in 1997 and targets for renewable energy generation, both of which contributed to wind energy growth. This system differs from the system employed in Germany in that the wind project developer can choose between a set feed-in tariff and a price for electricity that is a certain amount above the market price.[208]

Annual installed wind capacity in Spain increased from 419 MW in 1998 to 3,522 MW in 2007, but fell to 1,609 MW in 2008. Spain ranks third in the world in cumulative installed wind capacity.[209]

Spanish companies and Vestas have the largest shares of the market in Spain. Gamesa accounted for 57 percent of all cumulative turbine installations through the end of 2007 and 47 percent of turbines installed in 2007 (figure 22).[210] Vestas accounted for 15 percent of all cumulative turbine installations in Spain through the end of 2007 and 20 percent of turbines installed in 2007. Newer market entrant Acciona has a lower historical market share, but accounted for over 19 percent of wind turbine installations in 2007. Spanish company Ecotecnia was fourth in 2007 with 4 percent of the market. Navantia-Siemens accounted for 4 percent of turbine installations, GE 3.1 percent, and Enercon 1.9 percent.[211] OEMs or suppliers often must have local production facilities in order to meet the local content requirements established by regional governments.[212]

North and South America

The growth in wind installations in Canada, Mexico, and South America has been sporadic, but these markets are growing and there are export opportunities for U.S. producers. Imports of wind turbines by Brazil, Canada, and Mexico increased from $32 million in 2003 to $747 million in 2008, with Canada accounting for 71 percent of this increase (figure 23).[213] Other countries may also be potential wind markets—as noted earlier, CTC/DeWind exported 10 wind turbines to Chile in late 2008/early 2009.

Canada was tenth in the world in wind turbine installations in 2008 and third in wind turbine imports. Annual wind turbine installations increased from 86 MW in 2003 to 526 MW in 2008. Installations in 2009 are expected to rise to 650 MW despite the financial crisis.[214] In 2008, the leading wind turbine suppliers were Siemens (33 percent), Vestas (31 percent), GE (30 percent), and Acciona (5 percent).[215] GE is not currently exporting wind turbines from the United States to Canada, but Acciona, as noted earlier, plans to build some turbines for wind projects in Canada in the United States.

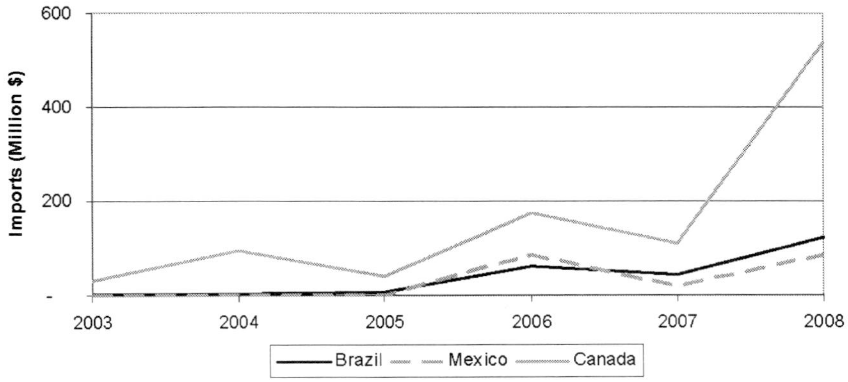

Source: GTIS, Global Trade Atlas (accessed various dates).

Figure 23. Imports of wind-powered generating sets, Brazil, Canada, and Mexico, 2003–08.

Canadian government policies will likely lead to further wind turbine installations—many provinces have established renewable energy goals (ranging between 500 MW and 4,600 MW) and the federal government provides an incentive for renewable energy generation.[216] In Canada, wind farms are generally developed as a result of requests for proposals by provincial utilities. While this can lead to large contracts, one government, Quebec, has local content requirements for its contracts that reduce the ability of companies outside Canada to compete for these contracts.[217] Despite some challenges, such as the need to upgrade transmission infrastructure, there is excellent potential for wind energy generation in Canada due to good wind resources and the ability to sell this electricity into both the Canadian and American markets.[218]

Brazil and Mexico both ranked in the top 20 in wind turbine imports in 2008.[219] 94 MW were installed in Brazil in 2008 and, while there were no reported installations in Mexico, 143 MW was built in Oaxaca in 2008 and expected to come online in 2009.[220] Markets in Brazil and Mexico are currently small in comparison to China, Europe, and the United States, but both countries have excellent wind resources and there is significant potential for market growth.[221]

REFERENCES

Articles/Reports/Presentationsn

Aarke, Dwight and Ron Haugen. *Wind Turbine Lease Considerations for Landowners.* North Dakota State University Extension Service. February 2009. http://www.ag.ndsu.edu (accessed April 7, 2009).

American Council on Renewable Energy. "Overview, Renewable Energy Provisions, American Recovery and Reinvestment Act of 2009." http://www.acore.org (accessed March 9, 2009).

American Wind Energy Association (AWEA). *AWEA 2008 Annual Rankings Report.* April 2008. http://www.awea.org (accessed September 24, 2008).

________*Annual Wind Industry Report: Year Ending 2008.* 2009. http://www.awea.org (accessed April 16, 2009).

________*3rd Quarter 2008 Market Report.* October 2008. http://www.awea.org (accessed January 31, 2008).

BTM Consult ApS. *International Wind Energy Development: World Market Update 2008.* March 2009.

Batten, Kit & Kari Manlove. (2008). *Identifying Hurdles to Renewable Electricity Construction.* Center for American Progress. December 18. http://www.americanprogress.org (accessed various dates).

Danish Wind Industry Association. "Vindmølleindustriens Branchestatistik 08." http://www.windpower.org (accessed various dates).

European Wind Energy Association (EWEA). *Wind at Work: Wind Energy and Job Creation in the EU.* January 2009. http://www.ewea.org (accessed February 2, 2009).

German Wind Energy Association. *Wind Energy-An Energy Source with a Fantastic Future.* March 2009. http://www.wind-energie.de (accessed January 29, 2009).

Global Wind Energy Council. *Global Wind 2008 Report*. May 2008. http://www.gwec.net (accessed various dates).

________*Global Wind 2007 Report*. May 2008. http://www.gwec.net (accessed various dates).

________*Global Wind 2006 Report*. http://www.gwec.net (accessed various dates).

International Energy Agency. *IEA Wind Energy 2007 Annual Report*. July 2008. http://www.ieawind.org (accessed various dates).

Johnson, Anna & Staffan Jacobson. (2002). "The Emergence of a Growth Industry: A Comparative Analysis of the German, Dutch and Swedish Wind Turbine Industries." Winter. http://www.druid.dk/conferences/winter2002/gallery/jacobsson.pdf (accessed January 29, 2008).

Lewis, Joanna I. (2007). "A Comparison of Wind Power Industry Development Strategies in Spain, India, and China." Prepared for the Center for Resource Solutions, July 19. http://www.resourcesolutions.org (accessed January 29, 2009).

________ "*A Review of the Potential International Trade Implications of Key Wind Power Industry Policies in China*." Paper prepared for the Energy Foundation China Sustainable Energy Program, October 2007. http://www.resource-solutions.org (accessed February 5, 2009).

________ "Technology Acquisition and Innovation in the Developing World: Wind Turbine Development in China and India." *Studies in Comparative International Development* 42, no. 3–4 (December 2007), 208–232.

Lewis, Joanna. & Ryan, Wiser. (2005). "A Review of International Experience with Policies to Promote Wind Power Industry Development." Prepared for the Energy Foundation China Sustainable Energy Program, March 10. http://www.resource-solutions.org (accessed January 29, 2009).

Li Jungfeng, Gao Hu, Shi Pengfei, Shi Jingli, Ma Lingjuan, Qin Haiyan, and Song Yanqin. *China Wind Power Report 2007*. Beijing: China Environmental Science Press.

Liebreich, Michael. (2008). "*Food for Thought-Abu Dhabi*." New Energy Finance, Presentation, November 6.

Magee, Joshua T. (2008). "Wind Turbine Supply Chain Trends and Component Tiers." Presentation at AWEA windpower supply chain workshop, Cleveland, OH, December 9.

Maves, Tom. (2008). "Wind Industry in Ohio and Supply Chain Opportunities." Ohio Department of Economic Development. http://www.development.ohio.gov (accessed March 5, 3009).

Merrill Lynch. (2007). *Wind Turbine Manufacturers; Here Comes Pricing Power*. August 10. http://www.ohiowind.org (February 4, 2008).

Navigant Consulting. (2008). *Economic Impacts of the Tax Credit Expiration*. Prepared for AWEA and the Solar Energy Research and Education Foundation, February 13. http://www.awea.org (accessed February 4, 2009).

North American Electric Reliability Council (NERC). (2008). *2008 Long-Term Reliability Assessment*. October. http://www.nerc.com (accessed various dates).

Peterson, Melissa. (2008). "Wind Energy Options for Farmers & Rural Landowners," PowerPoint presentation, June 27. http://www.cleanenergyresourceteams.org (accessed April 7, 2009).

Saint Consulting Group. *2009 Saint Index*. http://www.tscg.biz/survey/summary.html (accessed April 7, 2009).

Spanish Wind Energy Association. *Wind Power 2008, Sector's Yearbook: Analysis and Data.* http://www.aeeolica.es/ (accessed January 22, 2009).

Vestergaard, Jens. Lotte, Brandstrup. & Robert D. Goddard, III. (2004). "*Industry Formation and State Intervention: The Case of the Wind Turbine Industry in Denmark and the United States.*" Online version of a paper published in the Academy of International Business (Southeast USA Chapter) Conference Proceedings (November), 329–340. http://www.hha.dk/man paper2.pdf (accessed January 30, 2009).

Utility Wind Integration Group. (2006). *Utility Wind Integration State of the Art.* May. http://www.uwig.org/ (accessed May 20, 2009).

World Bank, International Bank for Reconstruction and Development. (2008). *International Trade and Climate Change: Economic, Legal, and Institutional Perspectives.* Washington, DC. http://web (accessed February 2, 2009).

Databases

APEC Tariff Database. http://www.apectariff.org (accessed December 17, 2008).

Bureau van Dijk. Orbis Companies Database. https://orbis.bvdep.com (accessed May 24, 2009). Global Trade Information Service, Inc. (GTIS). World Trade Atlas Database (accessed various dates). International Customs Tariffs Bureau. http://www.bitd.org (accessed December 17, 2008).

International Energy Agency (IEA). IEA Energy Technology R&D Statistics Service. http://www.iea.org/Textbase/stats/rd.asp (accessed various dates).

North Carolina Solar Center, NC State University. Database of State Incentives for Renewables and Efficiency. http://www.dsireusa.org (accessed various dates).

U.S. Department of Commerce (USDOC). U.S. Census Bureau (Census). Official U.S. trade statistics. http://www.census.gov/foreign-trade/download/dvd/index.html#merch (accessed monthly).

Government Documents (U.S.)

GE Energy Infrastructure. Edward C. Lowe. Written testimony submitted to the House Committee on Energy and Commerce, Subcommittee on Energy and Environment, hearing on Renewable Energy: Complimentary Policies for Climate Legislation, February 26, 2009.

U.S. Department of Energy, Office of Energy Efficiency and Renewable Energy. *20% Wind Energy by 2030: Increasing Wind Energy's Contribution to U.S. Electricity Supply*. June 2008. http://www1.eere.energy (accessed February 1, 2009).

U.S. Department of Energy, Energy Information Administration (EIA). *Capacity Additions, Retirements and Changes by Energy Source*. January 21, 2009. http://www.eia.doe.gov (accessed February 3, 2009).

U.S. Electric Net Summer Capacity, 2003–200 7. April 2009. http://www.eia.doe.gov (accessed June 4, 2009).

Harper, John P., Matthew, D. Karcher, & Mark Bolinger. (2007). *Wind Project Financing Structures: A Review & Comparative Analysis*. LBNL-63434. Lawrence Berkeley National Lab, Environmental Energies Technology Division, September. http://eetd.lbl.gov (accessed various dates).

Kelly, Marguerite. (2008). "Wind Energy Update." National Renewable Energy Lab, June 26.

Lewis, Joanna. & Ryan, Wiser. (2005). *Fostering a Renewable Energy Technology Industry: An International Comparison of Wind Industry Support Mechanisms*. LBNL-59 116. Lawrence Berkeley National Lab, Environmental Energies Technology Division, November. http://eetd.lbl.gov (accessed January 28, 2008).

U.S. International Trade Commission. Commission order issued August 30, 1996. http://www.usitc.gov (accessed various dates).

"ITC Institutes Section 337 Investigation on Certain Variable Speed Wind Turbines and Components Thereof." News release, March 25, 2008. http://www.usitc.gov (accessed December 16, 2008).

Certain Variable Speed Wind Turbines and Components Thereof. Investigation No. 337-TA-376.

Certain Variable Speed Wind Turbines and Components Thereof. Investigation No. 337-TA-641.

Wiser, Ryan. & Mark, Bolinger. (2008). *Annual Report on U.S. Wind Power Installation, Cost, and Performance Trends: 2007*. U.S. Department of Energy, Office of Energy Efficiency and Renewable Energy, May. http://www1.eere.energy (accessed September 22, 2008).

Wiser, Ryan. Mark, Bolinger. & Galen, Barbose. (2007). *Using the Federal Production Tax Credit to Build a Durable Market for Wind Power in the United States*. LBNL-63583. Lawrence Berkeley National Laboratory, November. http://eetd.lbl.gov (accessed various dates).

Government Documents (Non-U.S.)

German Federal Ministry for the Environment, Nature Conservation and Nuclear Safety. "Deployment of Renewable Energy Sources in Germany in 2007, Graphics and Tables." Based on statistical data from the Working Group on Renewable Energies-Statistics, December 15, 2008. http://www.bmu.de/english (accessed January 29, 2009).

Kjaergaard, Inge & Naadia, L. Kristensen. "Any Way the Wind Blows." *Zooming In: Expo 2008*, Ministry of Foreign Affairs of Denmark.

Industry Financial Reports

Clipper Windpower. *2007 Annual Report to Shareholders*. http://www.clipperwind.com (accessed October 15, 2008).

Clipper Windpower. *Annual Report and Financial Statements for the Year Ended December 31, 2008*. http://www.clipperwind.com (accessed various dates).

Composite Technology Corporation. *Form 10-K/A*. May 20, 2008.